Multi-Criteria Decision-Making

This textbook for advanced undergraduate and graduate students provides a clear and practical guide to the principles and methods for structuring and solving complex problems that include multiple, often conflicting criteria.

The book introduces the foundations of multi-criteria decision-making (MCDM) and its relationship with complementary approaches such as multi-objective optimization and goal programming. It then presents three major MCDM methods: reference-type methods such as TOPSIS, aggregation-type methods such as AHP and outranking-type methods such as ELECTRE. Each method is explained with numerical examples and strengths and limitations discussed. Preprocessing techniques are also covered. The book highlights user-friendly software tools, including Microsoft Excel-based programs and open-source Python libraries. End-of-chapter exercises are also provided to ensure that readers can consolidate their knowledge and apply it to real-world applications.

This textbook is vital for advanced undergraduate and graduate students taking decision science and optimization courses in disciplines such as engineering, business, finance and management. Its focus on recent methods and developments will ensure that students are well prepared for employment in their field or for further research alike.

Zhiyuan Wang is a lecturer at Singapore University of Social Sciences, where he teaches within the School of Business. Prior to this, he served as an assistant professor at the DigiPen Institute of Technology Singapore. He holds a B.Eng. and a Ph.D. from the National University of Singapore (NUS) and an M.Sc. from the Nanyang Technological University (NTU).

Gade Pandu Rangaiah has been with NUS since 1982, where he has received numerous teaching awards and is now an emeritus professor. Prior to joining NUS, he earned a B.Tech. from the Andhra University, India; an M.Tech. from the I.I.T. Kanpur, India; and a Ph.D. from the Monash University, Australia and worked at Engineers India Limited for two years.

Multi-Criteria Decision-Making

Principles, Methods and Programs

Zhiyuan Wang and Gade Pandu Rangaiah

CRC Press
Taylor & Francis Group
Boca Raton London New York

CRC Press is an imprint of the
Taylor & Francis Group, an **informa** business

Designed cover image: Getty Images

First published 2026
by CRC Press
2385 NW Executive Center Drive, Suite 320, Boca Raton FL 33431

and by CRC Press
4 Park Square, Milton Park, Abingdon, Oxon, OX14 4RN

CRC Press is an imprint of Taylor & Francis Group, LLC

ISBN: 978-1-032-85363-5 (hbk)
ISBN: 978-1-032-87010-6 (pbk)
ISBN: 978-1-003-53043-5 (ebk)

DOI: 10.1201/9781003530435

Typeset in Palatino
by Apex CoVantage, LLC

Contents

Preface

Decision-making is a central part of human activity. Whether selecting a field of study, evaluating investment opportunities, or designing complex engineering systems, we often face choices involving multiple alternatives, each assessed against many, often conflicting, criteria. The discipline of multi-criteria decision-making (MCDM) provides systematic tools and structured approaches to navigate such complexity. Over the past many decades, MCDM has matured into a vibrant and indispensable area of study, applied across engineering, business, finance, science, public policy and management.

This book has been written with the intention of offering both clarity and practical value. It begins with an introduction to the principles of MCDM and illustrates their relevance through examples drawn from daily life and professional practice. It then turns to multi-objective optimization (MOO) as a complementary discipline, which generates Pareto-optimal solutions and explains how MCDM and MOO work together in practice. Goal programming (GP) follows as another important technique pertinent to MCDM, showing how conflicting objectives can be managed by minimizing deviations from desired targets.

Before presenting MCDM methods, the book discusses some essential pre-processing steps, including normalization techniques for reconciling criteria measured on different scales, as well as both subjective and objective weighting methods for determining criteria weights. With this foundation in place, the book then presents three major types of MCDM methods in three separate chapters: reference-type methods such as technique for order of preference by similarity to ideal solution (TOPSIS) and VlseKriterijumska Optimizacija I Kompromisno Resenje (VIKOR), aggregation-type methods such as analytic hierarchy process (AHP) and simple additive weighting (SAW) and outranking-type methods such as elimination et choix traduisant la realité (ELECTRE) and preference ranking organization method for enrichment evaluations (PROMETHEE). Each method is introduced by describing its underlying principle and algorithm, followed by a clear numerical example that demonstrates the computational steps and concluded with a discussion of its advantages and disadvantages. This approach is followed consistently throughout the chapters.

The book also highlights software tools and programs that make MCDM methods accessible to readers for using them easily and correctly in applications. By judiciously blending theory with worked examples and computational resources, the book aims to bridge the gap between conceptual understanding and real-world problem solving.

The primary audience for this book includes advanced undergraduate and graduate students in engineering, business, finance, management, operations research and related disciplines. It is suitable as a main or supplementary textbook for courses in decision science and optimization. The structure of the chapters allows instructors to adapt the material for modular use as per their needs, while students can engage with the text independently to strengthen their learning. Exercises at the end of each chapter provide opportunities to apply and deepen the concepts discussed.

Beyond the classroom, the book can also be used by professionals seeking to compare different MCDM methods, choose software/programs, refine their analyses and adopt structured decision-making practices in their organizations. For researchers, the text offers a comprehensive foundation from which to explore advanced MCDM methods and their applications in diverse fields.

Writing this book has been motivated by the recognition that complex decisions require more than just intuition. They demand frameworks that are systematic, transparent and capable of incorporating multiple perspectives. Further, our research in MCDM over the last decade greatly facilitated the preparation of this book.

Although every effort has been made to ensure the accuracy of contents in this book, some errors or omissions may remain. We would greatly appreciate if readers could send their comments and corrections to help improve the book by writing to Dr. Wang (zywang@suss.edu.sg) and/or Prof. Rangaiah (chegpr@nus.edu.sg). Corrections will be addressed in future editions and updates will be shared from time to time through our Google Drive folder: https://drive.google.com/drive/folders/18ynIwnKaCRsptwBCrMIFVOlV9Mu_A7qM

Acknowledgments

I sincerely thank Prof. Rangaiah for his invaluable contributions to my research and this book. He has always been a beacon of wisdom in my career and life.

I am also grateful to the Singapore University of Social Sciences for supporting my teaching and research since I joined as a lecturer in 2025.

I am deeply grateful to my beloved wife, Xiaoyue, for her unwavering support during the preparation of this book, especially throughout her pregnancy. Her love gave me the courage to persevere through the most demanding moments of writing. To our daughter, 小月芽, I dedicate this book with all my love. May you grow up healthy and joyful, always keeping your childhood innocence and curiosity. Dad will do his utmost to protect you as you grow, nurture your dreams and support you always ~.

Zhiyuan Wang
Singapore

It has been my pleasure to supervise Zhiyuan's impactful research since 2016. His brilliance, commitment and diligence have been invaluable assets, culminating in coauthoring this book.

My heartfelt thanks to the National University of Singapore for supporting my research and teaching in process design, modeling, optimization and control for more than four decades.

I am profoundly grateful to my family for their unwavering support and deep affection. I am especially indebted to my dearest wife, Krishna Kumari. Her selfless care and devotion to both our family and me created the stable and nurturing foundation that allowed me to pursue my academic passions. Her support has been the silent, yet most significant, contributor to my achievements.

Gade Pandu Rangaiah
Singapore

1

Introduction

1.1 Overview

Multi-criteria decision-making (MCDM) refers to the process of making decisions in applications or situations involving many alternatives/choices/options (for selection, adoption, implementation, etc.), with each alternative assessed by multiple, often conflicting criteria (or objectives). These criteria characterize the performance, desirability, or benefits of available alternatives; one of these alternatives must be chosen by the decision-maker(s) after due consideration of trade-offs among the criteria values. MCDM has a wide range of applications in fields such as engineering, science, business, finance and public administration, as well as in everyday life. The following examples highlight the diversity of MCDM applications.

After completing (senior) high school, students have many options for further education; these options are for the course of study (e.g., accountancy, engineering, finance, law, management, medicine, or science) and the institution where they can study. Each combination of course and institution is one alternative for the student to choose and pursue. Criteria such as personal interest, career prospects, the reputation and location of the institution and costs influence this selection of one out of many alternatives. Multiple stakeholders (e.g., parents, siblings and close friends) besides the student often play a role in this decision-making. This MCDM problem is encountered and solved by millions of students worldwide each year. Hopefully, the decisions of the majority, if not all, of them are optimal considering the available alternatives, criteria and perspectives of decision-makers involved in each case.

Another example from personal life is choosing a family holiday destination. Families often have numerous alternatives (choices) for their vacation, ranging from local and regional to distant destinations. Criteria such as budget, travel time, safety, available activities and duration play a role in selecting the ideal holiday for a family. Typically, this decision involves input from multiple members of the family and such decision-making is required whenever a holiday is planned. The variety of chosen holiday destinations worldwide reflects the personalized (i.e., subjective) decision-making process of different families.

DOI: 10.1201/9781003530435-1

Purchasing a residential property is a major example of MCDM, which we often face a few times in our personal lives. Alternatives in this case involve properties in different suburbs and housing types such as apartments or (stand-alone) houses. Relevant criteria include location, budget, potential for value appreciation and the neighborhood reputation. (These criteria and/or their importance can change if the property is for investment.) Choosing a residential property is a significant decision often made collaboratively within a family. Millions of property transactions occur globally, emphasizing the importance of this complex decision-making process.

In finance, MCDM is frequently applied to the selection of investment portfolios, a complex decision faced by both individual investors and institutional fund managers. The alternatives in this case may include different combinations of asset classes such as stocks, bonds, mutual funds, exchange-traded funds (ETFs), or even alternative investments like real estate and commodities. Multiple, often conflicting, criteria influencing this decision include expected return, risk (e.g., volatility), liquidity, investment horizon, diversification potential and ethical considerations (e.g., environmental, social and governance factors). Depending on the investor's priorities and tolerance for risk, these criteria are weighted differently, resulting in distinct optimal portfolio choices for different decision-makers. The structured application of MCDM allows for a transparent and rational evaluation of investment alternatives, supporting informed financial decision-making in an environment of uncertainty and trade-offs.

In engineering, renewable energy source selection is one of the applications of MCDM. Selecting the most appropriate renewable energy source involves consideration of multiple criteria, such as efficiency, investment cost, operation and maintenance cost, land requirement, social acceptability and job creation. Alternatives in this scenario may include geothermal, solar, wind, hydroelectric and biomass energy sources. The systematic analysis of these alternatives based on multiple criteria demonstrates the utility of MCDM in addressing engineering challenges (Rani et al., 2020).

Many important chemical reactions require a catalyst and several catalysts are possible for a particular reaction. In such situations, the most preferred catalyst for the reaction under consideration must be chosen considering reaction conditions (of temperature and pressure), reaction rate enhancement, selectivity for the desired product, cost and environmental impact (in the production and/or disposal of used catalyst). Clearly, several alternatives and criteria are involved in the catalyst selection and so scientists need to employ MCDM in this application.

Another application of MCDM is in choosing the appropriate materials for manufacturing industrial power transmission gears for heavy loads (Dua et al., 2024), which are widely used in industrial machines, automobile gearboxes and construction vehicles. These gears are expected to operate in harsh conditions and endure high impact forces; therefore, the material used for them should have good durability, reliability and performance. The multifunctional aspects of these materials place restrictions and challenges on their selection. This decision-making process involves consideration of multiple criteria such as hardness, melting point, tensile strength, impact strength and cost.

MCDM is required for choosing programs or proposals out of many under consideration. For example, a large corporation can have a portfolio of strategic programs aimed at transforming various aspects of the business, ranging from operational efficiency to new product development. Then, it is required to determine whether to continue each of these programs (with or without adjustments) or terminate, based on a multi-criteria impact analysis. For this assessment, the criteria can be strategic alignment with the corporate goals, financial performance (e.g., return on investment), stakeholder and/or customer satisfaction, risk and issue severity, time to achieve benefits, synergy with other initiatives and social/environmental impact. Similar applications of MCDM are found in funding agencies, where it serves as a framework for evaluating and selecting research proposals for support from among many submissions, ensuring that decisions reflect a balanced consideration of multiple relevant criteria.

In social science, MCDM is often applied in policy evaluation and selection, where governments, nongovernmental organizations (NGOs), or research institutions must choose among various policy options to address societal challenges. Alternatives may include different programs or policies in areas such as education reform, poverty reduction, public health initiatives, or urban development strategies. Each option is assessed against multiple criteria, such as effectiveness, cost, social equity, public acceptance, feasibility of implementation and long-term sustainability. These criteria frequently conflict. For example, a policy that is highly effective may be costly or politically difficult to implement, requiring careful trade-offs by decision-makers. Applying MCDM in this context allows stakeholders to systematically weigh diverse priorities, balance competing interests and reach decisions that reflect both analytical rigor and social values.

In summary, each of the above examples involves alternatives with the performance of each of them characterized by several criteria. Owing to the conflicting nature of (some) criteria, the selection of the best or most preferred alternative requires thoughtful decision-making. While many MCDM problems in day-to-day life are resolved through intuition, experience and/or personal prioritization, systematic MCDM approaches for diverse applications in engineering, science, business and other domains offer consistency, reliability and transparency.

In the examples of MCDM outlined above, the most preferred alternative is likely to be different depending on the preference and acceptable trade-offs among conflicting criteria by the decision-maker(s) in each application/situation. Thus, there is no single right solution (i.e., the most preferred alternative) and subjectivity is involved in decision-making. Even then, the use of MCDM is desirable for comprehensive and consistent decision-making.

MCDM has been studied in various disciplines (Wang et al., 2024), including economics, business and industrial engineering. Partly because of this, the terminology associated with MCDM varies, with terms such as "criteria", "attributes" and "objectives" often used interchangeably. Consequently, MCDM is also termed multi-attribute decision-making (MADM), multi-criteria decision analysis/aid (MCDA) and multi-criteria analysis (MCA). MCDM problems are classified into two groups (Hwang and Yoon, 1981): MADM and multi-objective decision-making (MODM). Of these, the latter problems involve the formulation and solution of an optimization problem for multiple objectives (i.e., MOO), for example, to find the Pareto-optimal solutions, which are then used as alternatives in MCDM (Nabavi et al., 2024). An overview of MOO methods and applications in chemical engineering is available in Rangaiah et al. (2020). The book by Deb (2001) is a detailed resource for MOO using evolutionary algorithms.

The present book adopts the following terminology: criteria (synonymous with objectives and attributes), alternatives (synonymous with Pareto-optimal, non-dominated, non-inferior and efficient solutions) and alternatives–criteria matrix (ACM), which represents a table with alternatives listed in rows and criteria in columns. In the literature, ACM is known as a decision or objective matrix. Additionally, MCDM (often used interchangeably with MADM, MCDA and MCA) and MOO (synonymous with MODM) are employed. Depending on the application, MOO may be required to find the Pareto-optimal solutions (i.e., alternatives) before MCDM.

This Introduction chapter covers the general procedure of MCDM (in Section 1.2), people involved in MCDM (in Section 1.3), benefits of and challenges in MCDM (in Section 1.4) and the scope/use of this book (in Section 1.5). It ends with the summary in Section 1.6.

The two learning outcomes of this chapter are to (1) describe MCDM, its examples, procedure, persons involved, benefits and challenges and (2) outline the scope and use of this book.

1.2 General Procedure of MCDM

The general procedure of MCDM starts with the statement of the decision problem and consists of two major steps: problem formulation and solution (Figure 1.1). An example of the problem statement is to determine or find the most preferred renewable energy source, which is broad and open-ended and can have major significance for society. Another example is to choose the most preferred vendor for supplying materials and/or services to a company from a list of potential vendors; this example is specific and its effects are largely limited to the company. The problem statement leads to several questions such as possible alternatives and applicable criteria for evaluating

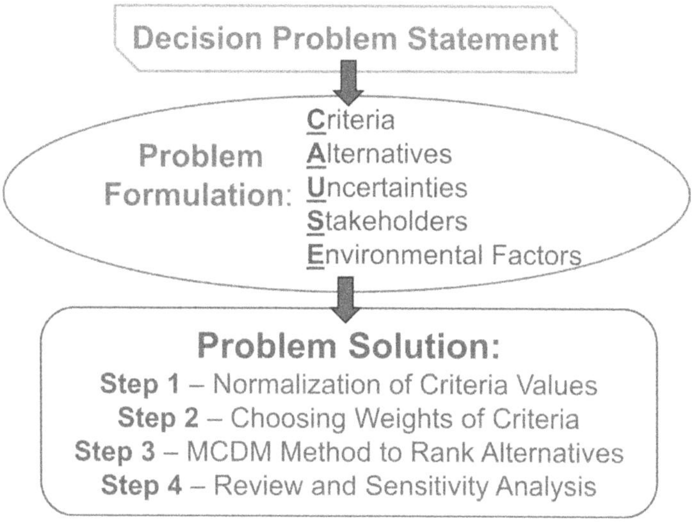

FIGURE 1.1
Main steps in MCDM.

each alternative and uncertainties. These are aptly captured by the acronym CAUSE (Belton and Stewart, 2002) that stands for criteria, alternatives, uncertainties, stakeholders and environmental factors (or constraints).

Details of CAUSE for two MCDM applications, reported in Pamučar and Ćirović (2015) and Lee and Rangaiah (2009), are summarized in Table 1.1. All these should be carefully and thoroughly explored and identified in the problem formulation for an MCDM application. In some applications, there can be a hierarchy of criteria (i.e., main and sub-criteria). For example, the location criterion in the scenario of purchasing a property can be divided into sub-criteria such as distance from work, distance from schools, distance from hospitals, distance from shops and distance from relatives/friends. On the other hand, in the recovery process design (Table 1.1), human toxicity potential, ecotoxicity potential, atmospheric potential and photochemical oxidation potential can be combined into a single criterion using suitable weights. This reduction in the number of criteria will simplify MCDM, but it may not provide a good understanding of sub-criteria and their variation. Further, MCDM results will depend on how sub-criteria are combined (e.g., weights assumed).

For developing CAUSE details, one can follow the sequence of criteria, alternatives, etc., as given in Table 1.1; however, some iteration may be necessary (e.g., returning to review the criteria after identifying alternatives and/or environmental factors) to ensure inclusion of all applicable CAUSE. It is also important to avoid an alternative that is unavailable or nonexistent. All these can affect the solution of the MCDM problem.

TABLE 1.1

CAUSE for Two MCDM Applications.

Factor	Application: Forklift Selection (Pamučar and Ćirović, 2015)	Application: Recovery Process Design (Lee and Rangaiah, 2009)
Problem statement	Select the most preferred forklift	Optimal design of a process for recovering volatile organic components
Criteria	Purchase cost, maintenance cost, fuel consumption, *warranty period, *service network, *availability of spare parts, *maximum load capacity, *maximum lifting capacity, *movement speed and *speed of lifting/lowering loads	Human toxicity potential, ecotoxicity potential, atmospheric potential, photochemical oxidation potential and *net present value
Alternatives	Seven different forklifts available in the market	More than 70 Pareto-optimal solutions found by optimizing the recovery process for multiple criteria simultaneously
Uncertainties	Variation in criteria (e.g., maintenance cost and fuel consumption) and environmental conditions over the life cycle of forklifts (Fuc et al., 2016)	Fluctuation in the feed flow rate and composition during the operation of the process
Stakeholders	Company purchasing and operators of the forklift	Company investing in the plant, engineers designing and operating the plant and society
Environmental factors (constraints)	Budget for forklift	Budget for setting up the recovery process

Note: The criteria beginning with * are the maximization (benefit) type, whereas all others (without *) are the minimization (cost) type

After preparing the CAUSE details, reliable values of all criteria for every alternative, uncertainties and environmental factors should be established in the problem formulation. These values may have to be found by suitable surveys (of markets, available products, technical literature and/or experts), technical analysis/research and/or optimization. For example, criteria values in the forklift application (Table 1.1) can be found by market and expert surveys, whereas criteria values in the recovery process design require the recovery process optimization for multiple objectives (i.e., MOO). Note that MOO itself involves two main steps: problem formulation (i.e., model development and validation) and solution (using a suitable optimization technique). MOO methods and programs for solving MOO problems and a comparison of MOO and MCDM are covered in Chapter 2.

Values of all criteria and alternatives can be expressed in an ACM, which is essentially a table of values for m alternatives in m rows and n criteria in n columns. This completes the step of problem formulation in MCDM, which is termed problem structuring and model building (i.e., two steps in Belton and Stewart, 2002). The problem formulation or model development is a critical step that requires extensive background and expertise related to the application under consideration. It may require substantial effort and time from several experts. At the end of problem formulation, it is extremely important to ensure that CAUSE and ACM are complete, correct and reliable (e.g., all possible alternatives are included, nonexistent or impossible alternatives are not included and ACM values are free from errors and uncertainties) for the next step (i.e., problem solution). Otherwise, results obtained from the problem solution can be unreliable.

Obviously, the next step after problem formulation is the problem solution. The key steps in the solution of MCDM problems are as follows:

Step 1: Normalization of Criteria Values – Each criterion in an MCDM application can have different magnitudes for one reason or another. Further, values of some criteria can be expressed in different dimensions (such as kg or ton and thousands of $ or millions of $), which affect the magnitude of the criteria values. To ensure uniformity across diverse criteria, values of each criterion are often normalized to the range of 0–1 (Wang et al., 2020). Several normalization methods are available in the literature; some popular normalization methods (e.g., max, max–min, sum and vector methods) are described in Chapter 4.

Step 2: Choosing Weights of Criteria – Many MCDM methods used in a later step require weights, which signify the relative importance of criteria or the acceptable trade-off between criteria. Decision-makers can provide weights. Alternatively, many methods are available for determining the weights of criteria (Wang et al., 2023). Some popular weighting methods (e.g., entropy, standard deviation (SD) and statistical variance methods) are covered in Chapters 5 and 6.

Step 3: Methods for Ranking Alternatives – These are often referred to as MCDM methods. A typical MCDM method ranks alternatives based on certain principles and using criteria weights and normalization (Wang et al., 2022). There are more than 200 MCDM methods for this ranking and new methods continue to be proposed in the literature. These methods can be broadly categorized into three main groups: reference-type, aggregation-type and outranking-type methods. Chapters 7–9 provide detailed descriptions of several popular and/or recent methods within each category of MCDM.

Step 4: Review and Sensitivity Analysis – It is essential to conduct sensitivity analysis of rankings to account for potential changes in weights and/or ACM (e.g., transformation of a criterion from maximization to minimization or vice versa and addition or deletion of one or more alternatives). See Nabavi et al. (2023) and Wang et al. (2021) for different approaches used in sensitivity analysis and Więckowski and Salabun (2023) for a thorough review of sensitivity analysis approaches in MCDM. Finally, the decision-maker(s) can review the top-ranked alternatives by one or more MCDM methods and the results of sensitivity analysis, with the aim of choosing one alternative for implementation.

1.3 People Involved in MCDM

Decision-makers involved in MCDM vary depending on the scope and significance of the application. For applications such as those in Table 1.1, there are likely to be one or more senior engineers, scientists, or managers from the relevant technical department in the company. For MCDM applications such as the location of a plant or warehouse, decision-makers will be the top management or board of the company. For large infrastructure applications of MCDM (such as improving the public transport in a city and choosing a renewable energy source) for the benefit of a larger society or region, decision-makers are likely to be elected members by the public (e.g., ministers), supported by their expert advisors.

In general, decision-makers in MCDM applications are unlikely to have sufficient background in the science and art of MCDM. Hence, besides them, several other people will be involved in MCDM. These include one or more facilitators, analysts and engineers/scientists, whose role is to provide technical support to decision-makers, who have the ultimate responsibility of making an informed decision. As the term implies, a facilitator plans and manages the entire process of MCDM application (e.g., understanding the problem and MCDM procedure including its variations and limitations, identifying criteria and alternatives, finding ACM and problem solution and managing the group of people involved in MCDM) to help decision-makers find the most preferred alternative.

The analysts are those with deep knowledge and expertise in MCDM methods and their software; they support decision-makers in performing the computations and interpreting the results. Finally and importantly, engineers and scientists provide the domain expertise in finding the alternatives and values of criteria in the MCDM application (through surveys of the market or literature and/or formulating and solving MOO problems to find the Pareto-optimal solutions, as in the recovery process design in Table 1.1). In the forklift application (Table 1.1), domain expertise of engineers is required for sourcing and preparing the list of vendors and models of forklifts as well as the values of the criteria.

Another group of people is the stakeholders, who may be affected by the decision made for the application, either positively or negatively. In MCDM applications such as forklift selection, all stakeholders (e.g., those using forklifts and shareholders of the company) are likely to have similar priorities of criteria and trade-offs. However, there may be diverse groups of stakeholders in large MCDM applications such as improving the road network in a city (whose stakeholders can be those using public transport, those using cars and those using bicycles) and setting up a plant in a natural habitat (whose stakeholders can be the company investing in the plant, people currently using the natural habitat and environmentalists). This makes the complex MCDM application even more challenging.

1.4 Benefits of and Challenges in MCDM

As stated earlier, MCDM involves subjectivity and does not give a unique answer for ranking the alternatives and/or selecting one of the alternatives. So, is MCDM beneficial? Should it be used for decision-making? The key benefits of MCDM are as follows: First, MCDM provides a systematic and thorough procedure for formulating and solving the decision problem. Second, the comprehensive procedure helps to develop a deeper understanding of the decision problem and various issues involved. The third benefit is the transparency and justifiability of the decision process to stakeholders, as and when required. Fourth, deliberations in MCDM can contribute to consensus building among decision-makers and other stakeholders. Fifth and finally, MCDM leads to informed and unbiased decisions instead of choosing an alternative based on intuition, majority preference, etc. Further, Keeney and Raiffa (1993) provide an in-depth discussion on how structured decision-making approaches, such as MCDM, help decision-makers systematically evaluate trade-offs and achieve clarity in complex scenarios. Hence, MCDM should be used in relevant applications.

To reinforce the above, readers are referred to Section 9.8 in Belton and Stewart (2002), where they share their experiences from several MCDM applications. These experiences demonstrate that MCDM can find a better alternative compared to that based on the intuition of decision-makers, lead to one or more new and better alternatives and/or confirm the preferred alternative based on decision-makers' intuition. Referring to the last point on confirmation, going through MCDM systematically is beneficial for providing a thorough understanding of the problem and issues involved by all decision-makers as well as the rationale and greater confidence in the chosen alternative.

To achieve the benefits of MCDM, readers and users of MCDM should be aware of the challenges that can affect the outcome. First and foremost is to ensure all applicable criteria, potential alternatives and environmental factors are included in the problem formulation. Then, the criteria values and uncertainties in them should be found as accurately and reliably as possible. Another issue is how to take care of uncertainty in the criteria values. If qualitative criteria are involved, how should the linguistic terms be converted into quantitative values for computations? If a main criterion has several sub-criteria, should all sub-criteria be considered individually or integrated as one main criterion? The values of some criteria and/or environmental factors can change over time. How should each of these be accounted for? There are studies on and techniques for addressing all these issues. The analysts involved in an MCDM project should have sound knowledge of these issues and potential solutions to them.

The recent review paper by Sahoo and Goswami (2023) highlights several challenges in MCDM. One issue is the subjectivity in weight assignments, as the result of MCDM heavily relies on the criteria weights provided by decision-makers. These weights are often subjective, introducing potential biases into the decision-making process. Using objective automated weighting techniques, such as entropy-based methods, can mitigate subjectivity. Another challenge is managing uncertainty and subjectivity, which are common in real-world scenarios due to the presence of imprecise or missing information. Fuzzy logic can be employed to handle ambiguity by allowing for flexible decision-making under uncertain conditions. Additionally, computational complexity and scalability pose barriers when dealing with large datasets or a high number of alternatives. The integration of MCDM with emerging technologies (Wang et al., 2025), including machine learning, AI and big data analytics, remains an underexplored yet crucial area.

In general, research continues to be pursued to address the above challenges in MCDM, to analyze MCDM methods, to apply MCDM to diverse applications, to expand MCDM to emerging areas, etc. To quantify the extent of research related to MCDM, the Scopus database was searched for publications with phrases of "Multi-Criteria Decision Making" or "Multicriteria Decision Making" or "Multi-Attribute Decision Making" or "Multiattribute Decision Making" or "Multi-Criteria Decision Analysis" or "Multicriteria Decision Analysis" in the title, abstract, or key words. The language of publications was limited to English.

Figure 1.2 displays the annual number of publications (as journal or conference papers, or reviews, books, book chapters, etc.) related to MCDM, as retrieved from the above search. The plot on the left shows that the annual number of publications in all subject areas increased from less than 10 to nearly 100 (about 10-fold) over the period 1980–2000. Then, the annual number of publications increased from less than 100 to nearly 5000 (about 50 times) from the year 2000 to 2024 (right plot of Figure 1.2). In case the search is limited to engineering (as categorized in Scopus), the annual number of publications related to MCDM shows similar trends as the publications in all subject areas (as included in Scopus). However, the former number (e.g., from around 30 in the year 2000 to close to 2000 in the year 2024 in the right plot of Figure 1.2) is about one-third of the number of publications in all subject areas. This confirms the importance of MCDM and its research and development in engineering and science.

For comparing research on MCDM with that on MOO, the Scopus database was searched for publications with phrases of "Multi-Objective Optimization" or "Multiobjective Optimization" or "Bi-Objective Optimization" or "Biobjective Optimization" or "Tri-Objective Optimization" or "Triobjective Optimization" in the title, abstract, or key words. The language of publications was limited to English. These search results are depicted in Figure 1.3. Comparison of this figure with Figure 1.2 shows that MOO-related publications are more than twice those on MCDM.

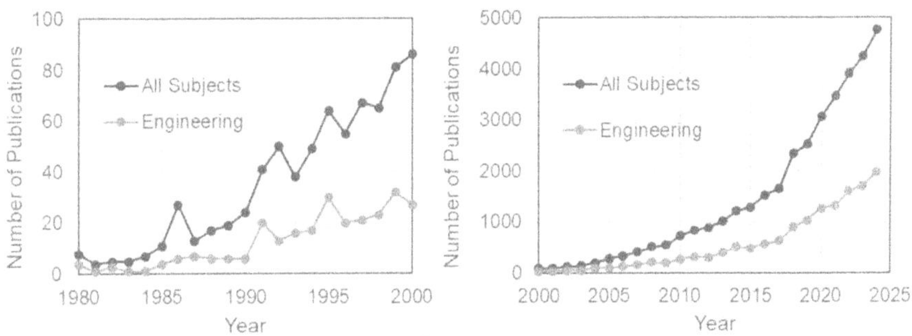

FIGURE 1.2
Yearly publications on MCDM: from 1980 to 2000 (left plot) and 2000 to 2024 (right plot). The y-axis scale is different in the two plots.

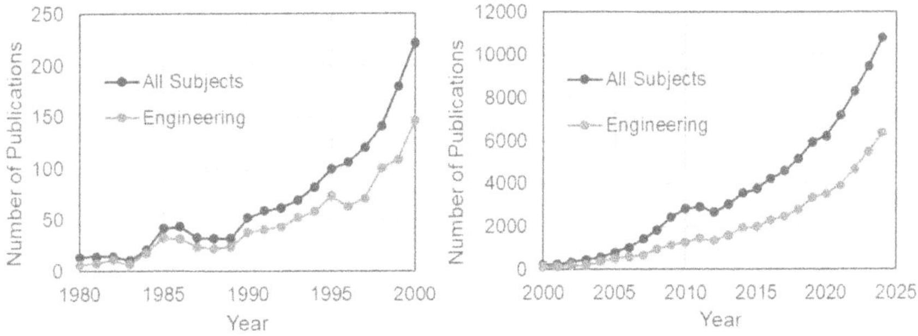

FIGURE 1.3
Yearly publications on MOO: from 1980 to 2000 (left plot) and 2000 to 2024 (right plot). The y-axis scale is different in the two plots.

Readers interested in learning about the recent developments and/or applications of MCDM or MOO in a particular area are advised to identify and read suitable review papers. For example, some review papers already cited in this chapter are Rangaiah et al. (2020) on MOO in chemical engineering, Sahoo and Goswami (2023) on MCDM advances and challenges and Więckowski and Salabun (2023) on sensitivity analysis in MCDM.

1.5 Scope and Use of This Book

As described earlier, the entire MCDM procedure consists of two major steps: problem formulation and problem solution. The problem formulation generally requires domain expertise related to the application. On the other hand, problem solution involves the selection and use of suitable MCDM and MOO methods (Nabavi et al., 2025), which are generic and can be employed in diverse applications. The focus of this book is on the common and/or recent methods for MCDM and MOO as well as easy-to-use programs/software for them.

Accordingly, this book primarily covers the principles, algorithms and relative merits of common and/or recent weighting, normalization and MCDM methods, which are all illustrated with suitable numerical examples (Figure 1.4). It also includes one chapter each on MOO (Chapter 2), GP (Chapter 3) and useful programs/software for MCDM (Chapter 10).

Chapter 2 is on the methods and programs for MOO. It introduces the concept of MOO, contrasts it with single-objective optimization (SOO) and explains Pareto-optimal solutions. Key methods such as weighted sum (WS), ε-constraint and metaheuristics (e.g., non-dominated sorting genetic

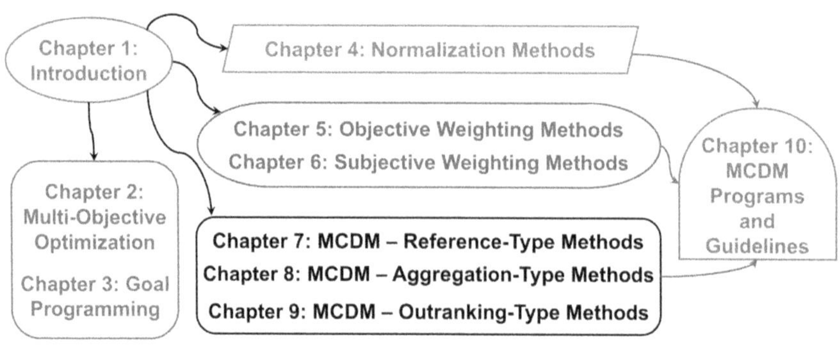

FIGURE 1.4
Chapters and their topics in this book.

algorithm II [NSGA-II]) are presented along with their strengths and limitations. This chapter also highlights software tools, including Excel-based EMOO, Python libraries: pymoo and SciPy and several MATLAB programs and explains how MOO and MCDM differ from one another while also showing how they can complement each other.

Chapter 3 covers GP, a widely used approach that bridges MOO and MCDM. It introduces the concept of setting target goals for each criterion and minimizing deviations from those goals, transforming optimization into a satisficing process. The chapter explains the three main GP variants, namely, weighted, lexicographic and min–max (Chebyshev) and demonstrates their application with examples. It also shows how GP can be applied to ACM data for ranking and selection. In addition, this chapter provides practical guidance on solving GP models using Excel Solver and Python PuLP. Finally, it summarizes GP's strengths, limitations and guidelines for effective use.

Chapter 4 focuses on normalization in MCDM. It explains why normalization of criteria values is essential when criteria differ in scale or units and presents four widely used methods: max, max–min, sum and vector. For each method, the chapter provides equations for benefit and cost criteria, numerical examples and a discussion of their behavior. It also addresses potential computational issues such as division by zero and handling negative values.

Chapter 5 describes seven objective weighting methods for MCDM: mean, SD, statistical variance, entropy, CRITIC, Gini coefficient and MEREC. These methods derive weights directly from ACM data without relying on decision-maker input. The chapter explains the principles, equations and examples for each method. **Chapter 6** presents subjective weighting methods for MCDM. It explains how these methods depend on decision-maker judgments rather than ACM data. Four key methods are covered: AHP, BWM, ROMs and SWARA, each illustrated with examples. The chapter compares their input needs and complexity, showing how subjective inputs shape the computed weights and offering guidance on when each method is most suitable.

MCDM/ranking methods can be broadly classified into three categories, namely, reference-type, aggregation-type and outranking-type methods. GP, described in Chapter 3, can be regarded as a special reference-type method in which the decision-maker provides criteria goals (collectively forming the reference solution). In addition to it, nine reference-type methods (TOPSIS, GRA, VIKOR, EDAS, MABAC, CODAS, PIV, MARCOS and PROBID) are presented in **Chapter 7**. Next, eight aggregation-type methods (SAW, MEW, AHP, ANP, COPRAS, MOORA, FUCA and WASPAS) and two prominent families of outranking-type MCDM methods (ELECTRE and PROMETHEE) are described in **Chapters 8 and 9**, respectively. In these chapters, the principles and algorithms of each MCDM method are first explained, followed by a clear numerical example illustrating the algorithmic steps. The merits and limitations of the method are then outlined. Each method discussed in these chapters can be read independent of the others.

Finally, **Chapter 10** focuses on readily available and easy-to-use programs/software for MCDM. It introduces EMCDM, an Excel-based tool; Python libraries: pymcdm and pyDecision; the R package RMCDA; and the desktop software SuperDecisions. These tools support various weighting and ranking methods and cater to users with different levels of technical experience. The chapter also mentions other commercial MCDM tools and provides guidelines for the reliable and robust MCDM process in practice. Selected applications of MCDM in engineering, science, business, finance and technology are given along with ACM for them in Appendix A.

This book will be valuable for undergraduate and postgraduate students, serving as either a primary or supplementary textbook for courses on MCDM, decision science and optimization. Students can reinforce their understanding by working through the exercises provided at the end of each chapter. The book will also benefit those who are interested in MCDM or engaged in its applications as decision-makers, facilitators, analysts, or practitioners. After reading the Introduction, each subsequent chapter can be read and understood independent of other chapters, by both students and professionals.

1.6 Summary

Main points in this Introduction chapter on MCDM are as follows:

1. MCDM is encountered in both personal and professional lives. Specifically, it occurs in diverse disciplines of engineering, science, business, economics and public administration.
2. MCDM facilitates systematic decision-making through in-depth analysis, deeper understanding and transparency to all stakeholders.
3. Yearly publications of studies and research on MCDM are about half of those on MOO, as per the search of the Scopus database.
4. MCDM consists of problem formulation and problem solution, both of which are equally important for correct and robust decision-making.
5. Solution of MCDM problems often requires normalization, weighting and ranking steps and many methods are available for each of these steps.
6. This book describes and illustrates the popular and/or recent methods for MOO, GP, normalization, weighting and MCDM/ranking methods, as well as readily available and easy-to-use programs/software.
7. This book is an excellent resource for students and practitioners interested in the broader MCDM field.

References

Belton, V. & Stewart T. J. (2002). *Multiple criteria decision analysis: An integrated approach.* Kluwer Academic Publishers.

Deb, K. (2001). *Multi-objective optimization using evolutionary algorithms.* John Wiley.

Dua, T. V., Duc, D. V., Bao, N. C., & Trung, D. D. (2024). Integration of objective weighting methods for criteria and MCDM methods: Application in material selection. *EUREKA: Physics and Engineering,* (2), 131–148. https://doi.org/10.21303/2461-4262.2024.003171

Fuc, P., Kurczewski, P., Lewandowska, A., Nowak, E., Selech, J., & Ziolkowski, A. (2016). An environmental life cycle assessment of forklift operation: A well-to-wheel analysis. *The International Journal of Life Cycle Assessment, 21,* 1438–1451.

Hwang, C.-L., & Yoon, K. (1981). *Multiple attribute decision making: Methods and applications.* Berlin: Springer-Verlag.

Keeney, R. L., & Raiffa, H. (1993). *Decisions with multiple objectives: Preferences and value trade-offs.* Cambridge University Press. https://doi.org/10.1017/cbo9781139174084

Lee, S. Q. E., & Rangaiah, G. P. (2009). Optimization of recovery processes for multiple economic and environmental objectives. *Industrial and Engineering Chemistry Research, 48,* 7662–7681.

Nabavi, S. R., Guo, Z., & Wang, Z. (2025). Machine learning-assisted surrogate modeling with multi-objective optimization and decision-making of a steam methane reforming reactor. *arXiv preprint arXiv:2507.07641.* https://doi.org/10.48550/arXiv.2507.07641

Nabavi, S. R., Wang, Z., & Rangaiah, G. P. (2023). Sensitivity analysis of multi-criteria decision-making methods for engineering applications. *Industrial & Engineering Chemistry Research, 62*(17), 6707–6722.

Nabavi, S. R., Wang, Z., & Rodríguez, M. L. (2024). Multi-objective optimization and multi-criteria decision-making approach to design a multi-tubular packed-bed membrane reactor in oxidative dehydrogenation of ethane. *Energy & Fuels, 39*(1), 491–503.

Pamučar, D., & Ćirović, G. (2015). The selection of transport and handling resources in logistics centers using Multi Attributive Border Approximation Area Comparison (MABAC). *Expert Systems with Applications, 42,* 3016.

Rangaiah, G. P., Feng, Z., & Hoadley, A. F. (2020). Multi-objective optimization applications in chemical process engineering: Tutorial and review. *Processes (Basel), 8*(5), 508. https://doi.org/10.3390/pr8050508

Rani, P., Mishra, A. R., Mardani, A., Cavallaro, F., Alrasheedi, M., & Alrashidi, A. (2020). A novel approach to extended fuzzy TOPSIS based on new divergence measures for renewable energy sources selection. *Journal of Cleaner Production, 257,* 120352.

Sahoo, S. K., & Goswami, S. S. (2023). A comprehensive review of Multiple Criteria Decision-Making (MCDM) methods: Advancements, applications and future directions. *Decision Making Advances, 1*(1), 25–48.

Wang, Y., Wang, Z., & Wu, Z. (2025). Multi-objective optimal control of nonlinear processes using reinforcement learning with adaptive weighting. *Computers & Chemical Engineering,* 109206.

Wang, Z., Li, J., Rangaiah, G. P., & Wu, Z. (2022). Machine learning aided multi-objective optimization and multi-criteria decision making: Framework and two applications in chemical engineering. *Computers & Chemical Engineering, 165*, 107945.

Wang, Z., Nabavi, S. R., & Rangaiah, G. P. (2023). Selected multi-criteria decision-making methods and their applications to product and system design. In *Optimization methods for product and system design* (pp. 107–138). Singapore: Springer Nature Singapore. https://doi.org/10.1007/978-981-99-1521-7_7

Wang, Z., Nabavi, S. R., & Rangaiah, G. P. (2024). Multi-criteria decision making in chemical and process engineering: Methods, progress and potential. *Processes, 12*(11), 2532. https://doi.org/10.3390/pr12112532

Wang, Z., Parhi, S. S., Rangaiah, G. P., & Jana, A. K. (2020). Analysis of weighting and selection methods for pareto-optimal solutions of multiobjective optimization in chemical engineering applications. *Industrial & Engineering Chemistry Research, 59*(33), 14850–14867.

Wang, Z., Rangaiah, G. P., & Wang, X. (2021). Preference ranking on the basis of ideal-average distance method for multi-criteria decision-making. *Industrial & Engineering Chemistry Research, 60*(30), 11216–11230.

Więckowski, J., & Salabun, W. (2023). Sensitivity analysis approaches in multi-criteria decision analysis: A systematic review. *Applied Soft Computing, 148*, 110915.

Exercises

E1.1 Like the examples in Section 1.1, outline one MCDM example in your area of interest. This should include as many details of CAUSE (Table 1.1) as possible.

E1.2 For the MCDM example outlined in the previous exercise, what are the ways to estimate the values of the criteria for the alternatives? Briefly discuss the challenges in estimating these values as well as possible uncertainties in them.

E1.3 Many MCDM problems involve trade-offs between conflicting criteria as well as some criteria that are correlated. Select an MCDM application in your area of interest and:

- Identify at least three conflicting criteria relevant to the application.
- Discuss how different stakeholders might prioritize these criteria differently.
- Identify two criteria in the application that may be correlated to some extent.

E1.4 State the people involved in MCDM and describe their respective roles in different stages of MCDM.

E1.5 For an MCDM example (e.g., your answer to E1.1 or any other), outline the benefits from the use of MCDM and potential challenges in performing MCDM.

2

Multi-Objective Optimization

2.1 Overview

Optimization is pervasive and beneficial in our lives and diverse disciplines of science, engineering and business, as well as many professions. In our everyday activities, be it daily routine, commuting, investing, playing sports, or planning holidays, we aim for the best or desired outcome. For instance, a daily commuter faces the problem of choosing an optimal route and/or mode of transport to work. A car might be the fastest option, but it comes with higher costs and the risk of traffic congestion. Public transportation, such as buses or trains, is often more affordable but may involve longer travel times and less flexibility. Generally, these activities are pursued optimally based on experience and qualitative reasoning, sometimes even without thinking and oftentimes without realizing the use of optimization qualitatively. The previous statement is applicable in professional activities too; for example, time management requires the optimal use of time. On the other hand, optimization in science, engineering, businesses and other professions generally refers to the utilization of quantitative models and computational methods for optimization.

Optimization for one objective (i.e., SOO) has been popular in many disciplines and professions for many decades. Effective computer programs for SOO are now readily available. However, optimization often involves more than one objective or performance criterion or goal, which necessitates the use of both MOO and MCDM. For example, production and/or distribution of a commercial product involves capital/operating costs; it also leads to some adverse environmental impact (e.g., carbon dioxide emissions). In this case, capital/operating costs and environmental impact must be minimized.

Quantitative optimization problems are formulated using appropriate models (i.e., mathematical equations based on fundamental principles and/or data-driven models) involving decision variables and constraints pertaining to the application (e.g., Nabavi et al., 2024). Solving an MOO problem through a quantitative or computational method gives a set of optimal solutions, which are equally good for the objectives considered

DOI: 10.1201/9781003530435-2

in the optimization problem. These solutions, known as Pareto-optimal or non-dominated or non-inferior solutions, represent trade-offs among the objectives (Wang et al., 2020). Out of these, ultimately only one optimal solution is required for implementation. This entails ranking the Pareto-optimal solutions by MCDM and then selecting, usually, the top-ranked solution. In this context, Pareto-optimal solutions from MOO are the alternatives in MCDM.

This chapter provides an overview of MOO problems, their solutions and popular computational methods for solving them. It also compares SOO with MOO and MOO with MCDM. Accordingly, the remaining sections in this chapter are as follows:

Section 2.2 Multi-Objective Optimization Problem and Its Solutions

Section 2.3 Comparison of Single- and Multi-Objective Optimization

Section 2.4 Methods for Solving Multi-Objective Optimization Problems

Section 2.5 Microsoft Excel, Python and MATLAB Programs for Multi-Objective Optimization

Section 2.6 Comparison between Multi-Objective Optimization and Multi-Criteria Decision-Making

Section 2.7 Summary

Good knowledge of the above contents is necessary for solving MOO problems and understanding their solutions. However, this is not required if the solutions or alternatives for MCDM of a particular application are available through domain knowledge and/or a survey of the literature/market relevant to the application.

The learning outcomes of this chapter on MOO are:

1. Describe the MOO problem and Pareto-optimal (non-dominated) solutions.
2. Outline popular methods and programs for solving MOO problems.
3. Select and apply a method and program for solving classical MOO problems.
4. State similarities and differences between MOO and SOO as well as between MOO and MCDM.

The presentation in this chapter assumes that readers have good background in SOO. If required, interested readers can refer to any textbook on optimization. For example, Rao (2009) covers optimization in detail, whereas the recent book by Ramteke et al. (2021) is comprehensive with good coverage of MOO and metaheuristic algorithms (MHAs).

2.2 Multi-Objective Optimization Problem and Its Solution

A general MOO problem involves (1) two or more objectives or performance criteria (to be either minimized or maximized, Eq. 2.1a), (2) a number of continuous decision variables (**x**), (3) one or more discrete decision variables (**y**), (4) lower and/or upper bounds (indicated by subscripts LB and UB) on continuous and discrete decision variables (Eq. 2.1b), (5) equality constraints (Eq. 2.1c) and/or (6) inequality constraints (Eq. 2.1d). Its mathematical form is as follows:

$$\text{Minimize} \qquad f_1(\mathbf{x},\mathbf{y}),\ f_2(\mathbf{x},\mathbf{y}),\ldots f_n(\mathbf{x},\mathbf{y}) \tag{2.1a}$$

$$\text{With respect to} \qquad \mathbf{x},\mathbf{y}$$

$$\text{Subject to} \qquad \mathbf{x}_{LB} \le \mathbf{x} \le \mathbf{x}_{UB};\ \mathbf{y}_{LB} \le \mathbf{y} \le \mathbf{y}_{UB} \tag{2.1b}$$

$$\mathbf{h}(\mathbf{x},\mathbf{y}) = \mathbf{0} \tag{2.1c}$$

$$\mathbf{g}(\mathbf{x},\mathbf{y}) \ge \mathbf{0} \tag{2.1d}$$

In Eq. 2.1a, there are n objectives; for simplicity, these are assumed to be minimization type. If some of the objectives are maximization type, then they can easily be transformed to minimization type by pre-multiplying each of such objectives by −1. Hence, the above MOO problem is generic. In Eq. 2.1, symbols in bold are vectors (e.g., several continuous/discrete decision variables and several equality/inequality constraints).

In Eq. 2.1d, the inequality constraints are shown as "greater than or equal to" (≥) type; they can also be of the "less than or equal to" (≤) type. If needed, constraints of the latter type can be converted to the former by pre-multiplying by −1. In real-world optimization problems, there may be only continuous or discrete variables, only bounds on decision variables and equality and/or inequality constraints. In other words, many types of optimization problems are possible. However, in an MOO problem, there will be at least two objectives to be optimized simultaneously.

For illustration in this chapter, the following mathematical (benchmark) MOO problem, namely, the CONSTR problem (Deb et al., 2002), is employed.

$$\text{Minimize} \qquad f_1(x_1,x_2) = x_1 \text{ and } f_2(x_1,x_2) = (1+x_2)/x_1 \tag{2.2a}$$

$$\text{With respect to} \quad x_1, x_2$$

Subject to $$0.1 \le x_1 \le 1.0;\ 0.0 \le x_2 \le 5.0 \quad (2.2b)$$

$$g_1(x_1, x_2) = 9x_1 + x_2 \ge 6.0 \quad (2.2c)$$

$$g_2(x_1, x_2) = 9x_1 - x_2 \ge 1.0 \quad (2.2d)$$

The above optimization problem has two objectives (both to be minimized simultaneously, Eq. 2.2a), two continuous decision variables with bounds (Eq. 2.2b) and two inequality constraints (Eqs. 2.2c and 2.2d); it has no discrete decision variables and no equality constraints.

Several sets of values of two objectives in Eq. 2.2a are displayed in Figure 2.1; this figure does not show values of decision variables and constraints. In other words, Figure 2.1 is a plot of solutions in the objective space only. All the points (solutions) in this figure satisfy the constraints (although not shown) and so they are feasible. Many feasible solutions are possible and only some of them are shown in Figure 2.1. Each of the two square points in this figure is inferior to the nearby diamond point in terms of both the objective values. That is, *between the square point and its nearby diamond point*, the latter is a non-dominated or non-inferior solution.

Among the diamond points, there is a clear trade-off between any two of them (i.e., a decrease in one objective is accompanied by an increase in the other objective). Comparing all the points in Figure 2.1, circle points are

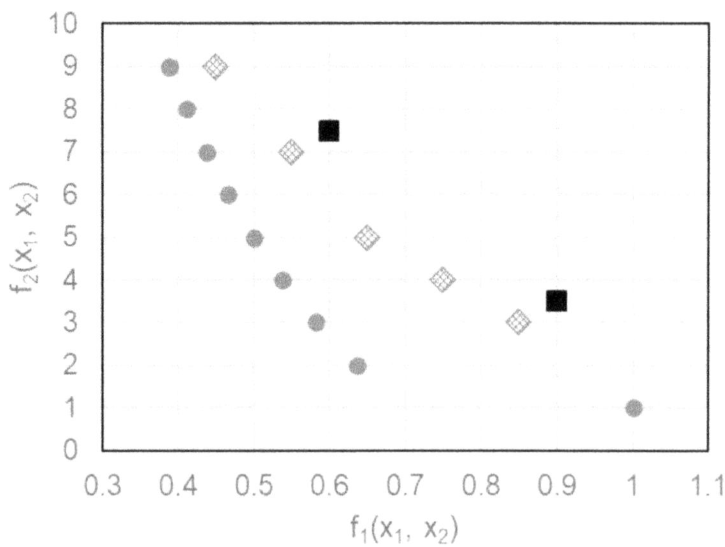

FIGURE 2.1
Non-dominated solutions (circles) and dominated solutions (diamonds and squares) of the CONSTR problem.

better than diamond and square points in one or both objectives; overall, circle points are non-dominated or Pareto-optimal solutions. Objective values of none of these solutions can be improved simultaneously; in other words, when the objective values of any two of these solutions are compared, one objective improves, whereas another worsens.

The trade-off among non-dominated solutions as well as some other solutions (e.g., diamond and square points in Figure 2.1) is due to conflict between the two objectives, which is often the case. For example, this conflict exists between maximizing profit and minimizing environmental impact in many manufacturing processes; this is because both these objectives increase with the raw material (and energy) used.

2.3 Comparison of Single- and Multi-Objective Optimization

Both SOO and MOO are similar in some respects, namely, decision variables and their nature (continuous or discrete), constraints and their types of equations (equality or inequality and linear or nonlinear). There are also similarities in the methods for solving SOO and MOO problems, namely, deterministic techniques such as the quasi-Newton method, successive quadratic programming (SQP) and the generalized reduced gradient (GRG) method and MHAs such as genetic algorithms (GAs) and particle swarm optimization (PSO); more on this is covered in the next section.

The main difference between SOO and MOO problems is in the number of objectives and the conflict (trade-off) among them. In the absence of any trade-off among the objectives in an MOO problem, a single set of values of decision variables will optimize all the objectives simultaneously. The main difference in the number of objectives with conflict leads to several differences between SOO and MOO problems, which are outlined below.

First, methods for solving SOO problems cannot be used for solving MOO problems; either the MOO problem should be somehow modified to an SOO problem, or the method for solving SOO problems should be adapted for handling multiple objectives. Second, the difficulty and computational effort for solving the optimization problem increase with the number of objectives (starting from the SOO problem). Third, an SOO problem typically has only one or a few optimal solutions (i.e., local and global optima), whereas an MOO problem has many non-dominated solutions. Note that some MOO problems have local and global Pareto-optimal fronts. Fourth, selection of an optimal solution for the SOO problem is trivial because there is only one (global) optimal solution, or the global optimal solution is obviously better. On the other hand, the selection of one of the Pareto-optimal solutions for implementation requires MCDM.

MOO addresses problems where multiple (conflicting) objectives must be optimized simultaneously. Instead of a single optimal solution, MOO seeks a set of equally good solutions, known as Pareto-optimal or non-dominated solutions, where no objective can be improved without degrading at least one other. The general procedure for solving MOO problems typically comprises (1) mathematical and/or data-driven model development and simulation, (2) formulation of the optimization problem, (3) solution of the optimization problem using a suitable computational method, (4) review and sensitivity analysis of the optimization results and (5) selection of one optimal solution (often integrated with MCDM here). See Rangaiah et al. (2020) for details on these steps. The following section focuses on computational methods for solving MOO problems (required in Step 3).

2.4 Methods for Solving Multi-Objective Optimization Problems

Many methods or techniques are available for solving MOO problems. They can be broadly classified into two categories, namely, single- and multi-objective approaches (Rangaiah et al., 2020), with several methods within each category. Available methods for solving MOO problems can also be classified in other ways based on whether many Pareto-optimal solutions are generated or not, whether inputs from decision-makers are required or not and when these inputs are needed (i.e., before, during or interactively and after problem solution); for more details, see Miettinen (1999) and Rangaiah (2017).

In the single-objective approach, the MOO problem is modified to an equivalent SOO problem, which can then be solved by the proven methods for SOO (Wang et al., 2024). Popular methods belonging to this approach are the weighted sum method (WSM), the ε-constraint method and goal programming (GP). These methods utilize the already available SOO methods and provide one non-dominated solution in each run or execution of the method and so many runs are required for finding many non-dominated solutions. In the multi-objective approach, the SOO methods developed for SOO are modified to handle multiple objectives. Methods in this category are MHAs such as GA, simulated annealing (SA), differential evolution (DE) and PSO. One single run of these methods provides numerous non-dominated solutions.

The WSM, ε-constraint method and MHAs are popular for solving MOO problems. Hence, these methods as well as their merits and limitations are presented in the following subsections. GP is the subject of the next chapter.

2.4.1 Weighted Sum Method

The WSM is probably the simplest and most intuitive method for solving MOO problems. It is based on combining all the objectives, each weighted with a suitable weight (w_i for the ith objective), to form a single objective. For example, n objectives in Eq. 2.1a are transformed into a single objective of weighted sum, WS of all objectives as follows:

$$\text{Minimize} \qquad WS(\mathbf{x}, \mathbf{y}) = \sum_{i=1}^{n} w_i f_i(\mathbf{x}, \mathbf{y}) \tag{2.3}$$

Here, w_i is between 0 and 1 and $\sum_{i=1}^{n} w_i = 1$. Decision variables and constraints are the same as those in the MOO problem (Eqs. 2.1b to 2.1d). The resulting SOO problem can be solved by an appropriate SOO method. The objectives in the MOO problem often have different magnitudes and so normalization of each objective to the range of, say, 0–1, is generally needed for formulating Eq. 2.3. This is illustrated in Example 2.1 below.

Example 2.1: Apply theWSM to solve the CONSTR problem in Eq. 2.2.

Solution: The WSM for solving the CONSTR problem is illustrated in several steps.

Step 1. For computing the range of objective values and normalization of objectives, solve the SOO problem, considering only one of the objectives in the MOO problem at a time. For the CONSTR problem, by taking $w_1 = 1$, the SOO problem will be $WS(x_1, x_2) = f_1(x_1, x_2) = x_1$. The decision variables and constraints remain the same as in Eqs. 2b–2d. This complete SOO problem is given below.

$$\text{Minimize} \qquad f_1(x_1, x_2) = x_1 \tag{2.4a}$$

$$\text{With respect to} \qquad x_1, x_2$$

$$\text{Subject to} \qquad 0.1 \le x_1 \le 1.0; 0.0 \le x_2 \le 5.0 \tag{2.4b}$$

$$g_1(x_1, x_2) = 9x_1 + x_2 \ge 6.0 \tag{2.4c}$$

$$g_2(x_1, x_2) = 9x_1 - x_2 \ge 1.0 \tag{2.4d}$$

This problem can be solved using the Solver tool in Microsoft Excel, which is easy to use (e.g., see the free online tutorial: https://www.solver.com/solver-tutorial-using-solver, accessed April 2025). Alternatively, Python libraries for solving SOO problems can be used. For instance, the pymoo library (Blank and Deb, 2020) implements a few optimization algorithms for SOO,

TABLE 2.1
Application of the Weighted Sum Method to Solving the CONSTR Problem.

Quantity	Min. $f_1(x_1, x_2)$	Min. $f_2(x_1, x_2)$	Min. $WS(x_1, x_2)$ with $w_1 = 0.3$	Min. $WS(x_1, x_2)$ with $w_1 = 0.5$	Min. $WS(x_1, x_2)$ with $w_1 = 0.7$
x_1	0.388889	1	0.666667	0.666667	0.478714
x_2	2.5	0	0	0	1.691578
$g_1(x_1, x_2)$	6	9	6	6	6
$g_2(x_1, x_2)$	1	9	6	6	2.616844
$f_1(x_1, x_2)$	0.388889	1	0.666667	0.666667	0.478714
$f_2(x_1, x_2)$	9	1	1.5	1.5	5.622523

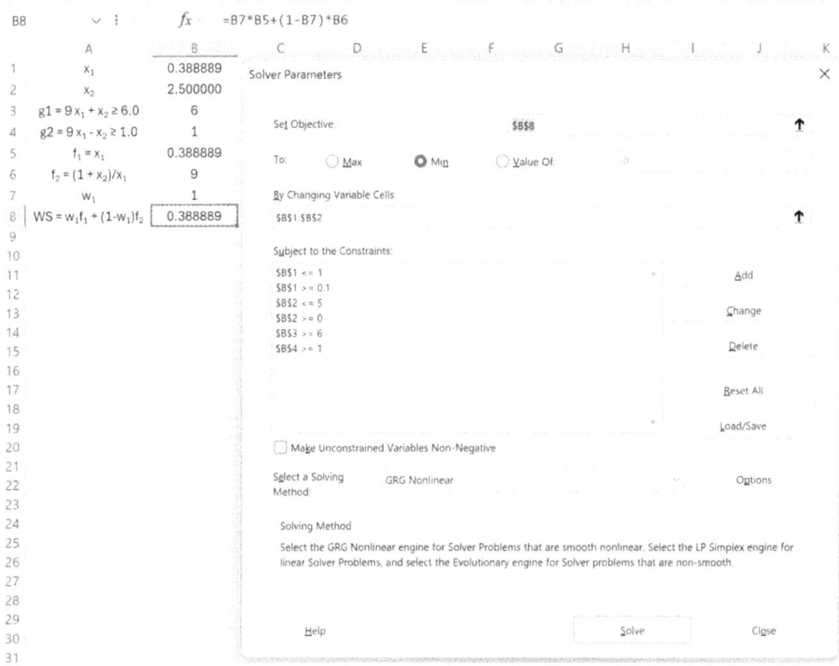

FIGURE 2.2
Excel Solver setup to solve the CONSTR problem using the weighted sum method when $w_1 = 1$.

such as Nelder–Mead simplex and pattern search techniques. Additionally, the SciPy library (Virtanen et al., 2020) implements a range of algorithms that can be employed for SOO, including the Broyden–Fletcher–Goldfarb–Shanno (BFGS) and the sequential least squares programming (SLSQP) method.

The solution of the above SOO problem using the Excel Solver tool is given in the second column of Table 2.1. The Solver setup is illustrated in Figure 2.2.

Step 2. Repeat Step 1 using the second objective, $f_2(x_1, x_2) = (1 + x_2)/x_1$ (instead of the first objective) in the MOO problem. The optimal solution for the SOO problem with the second objective is given in the third column of Table 2.1. Clearly, the range of $f_1(x_1, x_2)$ is $f_{1,min} = 0.3889$ to $f_{1,max} = 1.0$, whereas that of $f_2(x_1, x_2)$ is $f_{2,min} = 1.0$ to $f_{2,max} = 9.0$.

Step 3. Formulate the WS objective using the max–min normalization for each objective as follows: This normalization is typically needed because objectives in MOO problems can have different orders of magnitude. Without normalization, an objective with a larger magnitude would dominate the WS computation (as if this objective had an inherently large weight).

$$WS(x_1, x_2) = w_1\left(\frac{f_1(x_1, x_2) - f_{1,min}}{f_{1,max} - f_{1,min}}\right) + (1 - w_1)\left(\frac{f_2(x_1, x_2) - f_{2,min}}{f_{2,max} - f_{2,min}}\right) \quad (2.5)$$

Here, w_2 is taken as $1-w_1$ so that the sum of weights is unity. The solution to the SOO problem using the Excel Solver tool, with Eq. 2.5 as the objective, $w_1 = 0.3$, decision variables from Eq. 2.4b and constraints from Eqs. 2.4c and 2.4d, is shown in the fourth column of Table 2.1. Additionally, Python code for solving the same problem is provided in Table 2.2 and its output is displayed in Table 2.3.

Step 4. Repeat Step 3 with several values (> 0.0 and < 1.0) of w_1, as desired. Results for $w_1 = 0.5$ and 0.7 are shown in the last two columns of Table 2.1. Note that the WS minimization is equivalent to only minimizing $f_2(x_1, x_2)$ if $w_1 = 0.0$ and only minimizing $f_1(x_1, x_2)$ if $w_1 = 1.0$.

Comments: In Table 2.1, Pareto-optimal solutions are the same for $w_1 = 0.3$ and 0.5. This can occur for some values of w_1. In other words, the effect of weight on the SOO problem solution is not linear and so Pareto-optimal solutions found by the WSM may not be well distributed (e.g., like circles in Figure 2.1).

The main merit of the WSM is simplicity. Besides the need for an effective method for solving the resulting SOO problem, the method has several limitations. First, n weights satisfying $\sum_{i=1}^{n} w_i = 1$ must be chosen. Second, for each set of weights, only one Pareto-optimal solution is found and so the SOO problem, each time with different weights, must be solved multiple times to find many Pareto-optimal solutions. This becomes tedious if the number of objectives in the MOO problem is three or more. Third, it is difficult to find well-distributed Pareto-optimal solutions. Finally, the WSM fails to find the Pareto-optimal solutions in the non-convex region of the Pareto-optimal front (Kim and de Weck, 2005).

TABLE 2.2

Python Code Using the SciPy Library (version 1.13.1) for the Weighted Sum Method When $w_1 = 0.3$.

```
import numpy as np
from scipy.optimize import minimize

w1 = 0.3 # Weight for the first objective

# Define objective function (using weighted sum method)
def objective(x):
    f1 = x[0]
    f2 = (1 + x[1]) / x[0]

    # Use the next line of code for finding the range of objective values
    # return w1 * f1 + (1 - w1) * f2

    # Next line of code is for normalized f1 and f2
    return w1 * (f1 - 0.388889)/(1 - 0.388889) + (1 - w1) * (f2 - 1)/(9 - 1)

# Define constraints
def constraint1(x):
    # g1 >= 0
    return 9 * x[0] + x[1] - 6.0

def constraint2(x):
    # g2 >= 0
    return 9 * x[0] - x[1] - 1.0

# Lower and upper bounds for x1 and x2
bounds = [(0.1, 1.0), (0.0, 5.0)]

# Initial guess
x0 = np.array([0.1, 0.0])

# Define constraint type
constraints = [{'type': 'ineq', 'fun': constraint1},
               {'type': 'ineq', 'fun': constraint2}]

# Solve the SOO problem
solution = minimize(objective, x0, method='SLSQP',
                    bounds=bounds, constraints=constraints)

# Print results (up to 6 decimal places)
print(f"--------optimization results when w1 = {w1}--------")
print("x1:", np.round(solution.x[0], 6))
print("x2:", np.round(solution.x[1], 6))
print("g1:", np.round(9 * solution.x[0] + solution.x[1],6))
print("g2:", np.round(9 * solution.x[0] - solution.x[1],6))
print("f1:", np.round(solution.x[0], 6))
print("f2:", np.round((1 + solution.x[1])/solution.x[0], 6))
```

TABLE 2.3

Output of the Python Code in Table 2.2.

```
-------- Optimization results when w1 = 0.3 --------
x1: 0.666667
x2: 0.0
g1: 6.0
g2: 6.0
f1: 0.666667
f2: 1.5
```

2.4.2 ε-Constraint Method

The two key ideas in this method are to choose only one of the objectives in the MOO problem as the primary objective and to convert the remaining objectives into inequality constraints to formulate the equivalent SOO problem. Although the selection of the primary objective is somewhat arbitrary, it should make the SOO problem easy to solve (e.g., choose the strongly nonlinear objective in the MOO problem as the primary objective and avoid the addition of a strongly nonlinear constraint). The purpose of the new inequality constraints is to keep each of the remaining objectives (assumed to be minimization type) below some desired values. If the remaining objectives are of the maximization type, then they are kept above the desired values.

According to the ε-constraint method and retaining the last objective, $f_n(\mathbf{x}, \mathbf{y})$, as the primary or sole objective, the MOO problem in Eq. 2.1 is transformed to the following SOO problem.

Minimize $\quad f_n(\mathbf{x},\mathbf{y}) \quad$ (2.6a)

With respect to $\quad \mathbf{x},\mathbf{y}$

Subject to $\quad \mathbf{x}_{LB} \le \mathbf{x} \le \mathbf{x}_{UB};\mathbf{y}_{LB} \le \mathbf{y} \le \mathbf{y}_{UB} \quad$ (2.6b)

$$\mathbf{h}(\mathbf{x},\mathbf{y}) = \mathbf{0} \quad (2.6c)$$

$$\mathbf{g}(\mathbf{x},\mathbf{y}) \ge \mathbf{0} \quad (2.6d)$$

$$f_1(\mathbf{x},\mathbf{y}) \le \varepsilon_1 \quad (2.6e)$$

$$f_2(\mathbf{x},\mathbf{y}) \le \varepsilon_2 \quad (2.6f)$$

$$\ldots\ldots\ldots\ldots\ldots\ldots$$

$$f_{n-1}(\mathbf{x},\mathbf{y}) \le \varepsilon_{n-1} \quad (2.6g)$$

Notice that the resulting SOO problem has (n – 1) additional inequality constraints (Eqs. 2.6e–2.6g). Like the WSM, the ε-constraint method with one set of values for ε's gives only one Pareto-optimal solution.

Thus, both the WSM and the ε-constraint method transform multiple objectives (i.e., a vector) into a single objective (i.e., a scalar). Hence, they belong to the class of the scalarization approach.

Example 2.2: Apply the ε-constraint method to solve the CONSTR problem in Eq. 2.2.

Solution: The ε-constraint method is illustrated in several steps below.

Step 1. Let's select $f_1(x_1, x_2)$ as the primary objective. Then, $f_2(x_1, x_2)$ becomes an inequality constraint (Eq. 2.7e) in the following SOO problem. This constraint keeps $f_2(x_1, x_2)$ below some desired value (ε).

Minimize $$f_1(x_1, x_2) = x_1 \quad (2.7a)$$

With respect to $$x_1, x_2$$

Subject to $$0.1 \le x_1 \le 1.0; 0.0 \le x_2 \le 5.0 \quad (2.7b)$$

$$g_1(x_1, x_2) = 9x_1 + x_2 \ge 6.0 \quad (2.7c)$$

$$g_2(x_1, x_2) = 9x_1 - x_2 \ge 1.0 \quad (2.7d)$$

$$f_2(x_1, x_2) \le \varepsilon \quad (2.7e)$$

Step 2. For choosing suitable values of ε, perform Steps 1 and 2 (i.e., optimization considering only one objective at a time) of the WSM in Example 2.1. The obtained results in the second and third columns of Table 2.1 show that the range of $f_2(x_1, x_2)$ is $f_{2,min} = 1.0$ to $f_{2,max} = 9.0$. This means a suitable range of ε is 1.0–9.0.

Step 3. Solve the SOO problem in Eq. 2.7 using ε = 2.0. The results from this optimization using the Solver tool in Microsoft Excel are shown in the second column of Table 2.4. The Excel Solver setup is displayed in Figure 2.3.

Step 4. Repeat Step 3 for ε = 3.0, 4.0, 5.0, 6.0, 7.0 and 8.0 using the Solver tool in Microsoft Excel. The results from solving the corresponding SOO problem in Eq. 2.7 are presented in columns 3–8 of Table 2.4. Notice that ε = 1.0 and 9.0 are not considered because the solution of the SOO problem with these two values will be, respectively, the same as that in the second and third columns of Table 2.1. Additionally, Python code for solving the same problem is provided in Table 2.5 and its output is shown in Figure 2.4.

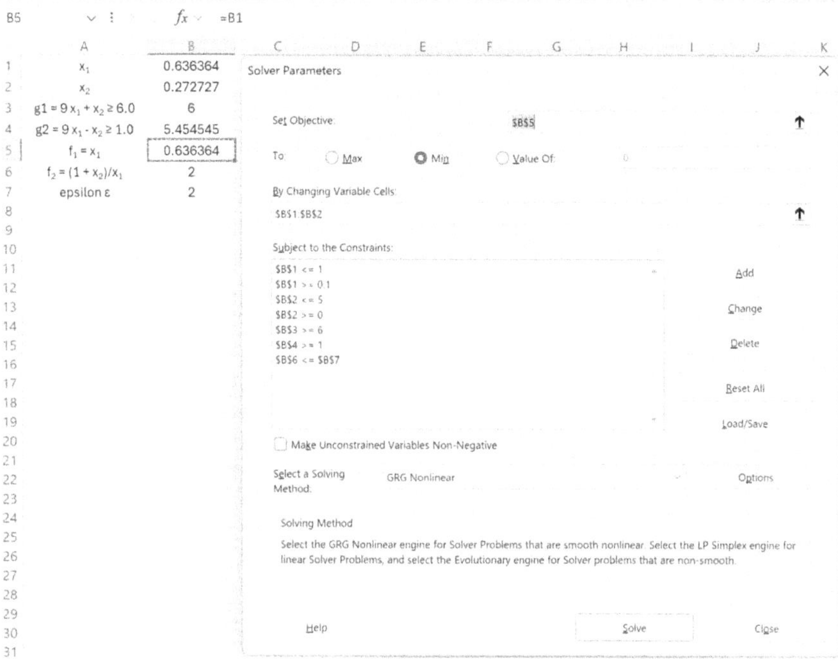

FIGURE 2.3
Excel Solver setup to solve the CONSTR problem using the ε-constraint method when ε = 2.

TABLE 2.4
Application of the ε-Constraint Method to Solving the CONSTR Problem.

Quantity	ε = 2.0	ε = 3.0	ε = 4.0	ε = 5.0	ε = 6.0	ε = 7.0	ε = 8.0
x_1	0.636364	0.583333	0.538462	0.5	0.466667	0.4375	0.411765
x_2	0.272727	0.75	1.153846	1.5	1.8	2.062499	2.294119
$g_1(x_1, x_2)$	6	6	6	6	6	6	6
$g_2(x_1, x_2)$	5.454545	4.5	3.692307	3.000001	2.4	1.875002	1.411762
$f_1(x_1, x_2)$	0.636364	0.583333	0.538462	0.5	0.466667	0.4375	0.411765
$f_2(x_1, x_2)$	2	3	4	4.999999	6.000001	6.999996	8.000006

Comments: In Table 2.4, the Pareto-optimal solution found for each ε value is different. The seven solutions in this table and those in the second and third columns of Table 2.1 are the circles in Figure 2.1. Clearly, solutions found by the ε-constraint method are evenly spread as per chosen ε values, particularly in terms of $f_2(x_1, x_2)$. However, there is a large gap in $f_1(x_1, x_2)$ between 0.636364 and 1.0.

TABLE 2.5

Python Code Using the pymoo Library (version 0.6.1.3) for the ε-Constraint Method.

```
import numpy as np
from pymoo.algorithms.soo.nonconvex.nelder import NelderMead
from pymoo.core.problem import ElementwiseProblem
from pymoo.optimize import minimize
import matplotlib.pyplot as plt

class CONSTR(ElementwiseProblem):

    def __init__(self, epsilon):
        super().__init__(n_var=2,
                         n_obj=1,
                         n_ieq_constr=3, # f2 becomes constr
                         xl=np.array([0.1, 0.0]), # lower bound
                         xu=np.array([1.0, 5.0])) # upper bound
        self.epsilon = epsilon

    def _evaluate(self, x, out, *args, **kwargs):
        f1 = x[0]

        # each constraint needs to be provided in the form of <= 0
        g1 = -9 * x[0] - x[1] + 6.0
        g2 = -9 * x[0] + x[1] + 1.0
        g3 = (1 + x[1]) / x[0] - self.epsilon # f2 becomes constr

        out["F"] = f1
        out["G"] = [g1, g2, g3]

# Range of ε values: from 1 to 9 (both inclusive)
epsilon_values = np.arange(1, 10, 1)

pareto_solutions = []

algorithm = NelderMead()

for epsilon in epsilon_values:
    problem = CONSTR(epsilon)

    res = minimize(problem,
                   algorithm,
                   seed=1,
                   verbose=False)

    pareto_solutions.append((res.X, res.X[0], (1 + res.X[1]) / res.X[0]))

# Print results
for i, (x, f1, f2) in enumerate(pareto_solutions):
    print(f"when ε = {epsilon_values[i]}: " \
          f"x1 = {x[0]:.6f}, x2 = {x[1]:.6f}, " \
          f"g1 = {9*x[0]+x[1]:.1f}, g2 = {9*x[0]-x[1]:.6f}, " \
          f"f1 = {f1:.6f}, f2 = {f2:.1f}")
```

TABLE 2.5 (***Continued***)

Python Code Using the pymoo Library (version 0.6.1.3) for the ε-Constraint Method.

```
#
Visualize Pareto-optimal front
plt.scatter([f[1] for f in pareto_solutions],
            [f[2] for f in pareto_solutions])
plt.xlabel("Objective function 1 (f1)")
plt.ylabel("Objective function 2 (f2)")
plt.show()
```

```
when ε = 1: x1 = 1.000000, x2 = 0.000000, g1 = 9.0, g2 = 9.000000, f1 = 1.000000, f2 = 1.0
when ε = 2: x1 = 0.636364, x2 = 0.272725, g1 = 6.0, g2 = 5.454554, f1 = 0.636364, f2 = 2.0
when ε = 3: x1 = 0.583335, x2 = 0.749990, g1 = 6.0, g2 = 4.500025, f1 = 0.583335, f2 = 3.0
when ε = 4: x1 = 0.538463, x2 = 1.153839, g1 = 6.0, g2 = 3.692326, f1 = 0.538463, f2 = 4.0
when ε = 5: x1 = 0.500001, x2 = 1.499997, g1 = 6.0, g2 = 3.000011, f1 = 0.500001, f2 = 5.0
when ε = 6: x1 = 0.466670, x2 = 1.799983, g1 = 6.0, g2 = 2.400048, f1 = 0.466670, f2 = 6.0
when ε = 7: x1 = 0.437501, x2 = 2.062500, g1 = 6.0, g2 = 1.875005, f1 = 0.437501, f2 = 7.0
when ε = 8: x1 = 0.411765, x2 = 2.294121, g1 = 6.0, g2 = 1.411766, f1 = 0.411765, f2 = 8.0
when ε = 9: x1 = 0.388902, x2 = 2.500049, g1 = 6.0, g2 = 1.000066, f1 = 0.388902, f2 = 9.0
```

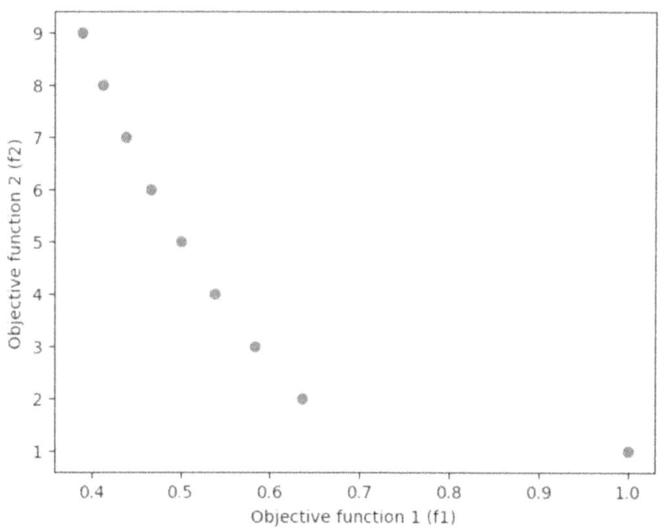

FIGURE 2.4
Output of the Python code for the ε-constraint method.

The CONSTR benchmark problem is simple with two objectives, two decision variables having bounds and two inequality constraints. It is relatively easy to solve by the WSM, ε-constraint method and MHAs (to be covered in the next section). Obviously, other MOO benchmarks and application problems are likely to be more challenging to solve.

The ε-constraint method has a few merits, namely, (1) its principle is easy to understand and (2) Pareto-optimal solutions are reasonably distributed as per chosen ε values; if not, the distribution of the solutions can easily be improved by solving the SOO problem for some more ε values. Its drawbacks are as follows: First, values of $\varepsilon_1, \varepsilon_2, \ldots, \varepsilon_{n-1}$ must be chosen. Second, for each set of ε's, only one Pareto-optimal solution is found; so, the SOO problem, each time with different ε's, must be solved multiple times to find many Pareto-optimal solutions. This becomes tedious if the number of objectives in the MOO problem is three or more. These two drawbacks are like those of the WSM. Third, the resulting SOO problem in the ε-constraint method is more difficult to solve due to the additional (n–1) inequality constraints.

2.4.3 Metaheuristic Algorithms (MHAs)

MHAs are iterative algorithms (like other methods/techniques) for solving SOO problems. The earliest of these were proposed and studied in the 1970s and 1980s. By the end of 2022, there were more than 500 MHAs, which are listed and briefly reviewed in Rajwar et al. (2023). "Heuristics" means rules based on common understanding and "metaheuristics" refers to high-level heuristics. Specifically, MHAs incorporate suitable operations and strategies for efficiently finding the global optimum (Wang et al., 2022). They use random numbers and their algorithms include both exploration strategies (for searching the entire search space and escaping from the local optimum) and exploitation strategies (for faster convergence to the optimum). Overall, MHAs can locate the global optimum in a reasonable computation time.

MHAs are inspired by natural and physical phenomena. Accordingly, they can be evolutionary algorithms (e.g., GA and DE), swarm intelligence algorithms (e.g., PSO and ant colony optimization [ACO]), physics-based algorithms (e.g., SA and harmony search) and miscellaneous (e.g., teaching learning-based algorithms) (Rajwar et al., 2023). Search for the optimum in many of these algorithms (e.g., GA, DE, PSO and ACO) is based on a population of solutions or points or agents, whereas search in some algorithms (e.g., SA and tabu search) is using only one solution/point at a time.

Almost all MHAs were initially developed for unconstrained SOO problems (with only bounds on decision variables). Then, (inequality) constraint handling techniques were incorporated into them for solving constrained SOO problems. Further, many MHAs were modified to handle multiple objectives for solving MOO problems. Generally, they generate many Pareto-optimal solutions in a single run or execution of the algorithm (e.g., Park et al., 2023).

The main steps in a typical MHA using a population for solving an MOO problem are as follows:

Step 1. Generate randomly an initial or parent population of solutions or individuals. Evaluate the objectives and constraints of each parent solution in this population.

Step 2. Systematically generate the child population using the parent population and random numbers. Evaluate the objectives and constraints of all individuals in the child population.

Step 3. Combine the parent and child populations. Find individuals that are better at satisfying constraints (i.e., feasible solutions) and objective values (considing non-dominance and crowding distance) for continuing to the next iteration (or generation).

Step 4. Repeat iterations from Step 2 until a termination criterion (e.g., improvement in non-dominated solutions in the recent iteration(s) and/or maximum number of iterations) is met.

For finding better individuals in Step 3, one approach that considers both feasibility and non-dominance of a solution is the constrained-dominance principle of Deb et al. (2002). According to this, solution A is a constraint-dominating solution B if any of the following conditions are met: (1) both solutions are not feasible, but overall constraint violation of A is less than that of B; (2) solution A is feasible, whereas solution B violates one or more constraints; and (3) both solutions are feasible and solution A dominates solution B in all objective values. The crowding distance mentioned in Step 3 is a measure of the distance of an individual solution to its nearest solutions in the objective space. It is employed to choose constraint-dominating solutions that are spread out in the objective space, for the next generation.

Obviously, there are many more details in the above Steps 2 and 3 in any MHA and there are variations in them from one MHA to another. For example, Deb (2001) covers many details on evolutionary algorithms including GAs for MOO; Sharma and Rangaiah (2013) describe an improved DE for MOO; and Chapter 4 in Ramteke et al. (2021) details GA, DE, SA and ACO for SOO and MOO. Interested readers can find more details in these sources.

The main merits of MHAs for MOO are as follows: First, they do not require continuity and derivatives of objectives and constraints; thus, they can solve any complex MOO problem. Second, MHAs do not require any initial estimates of the optimal solution. Third, they can find many Pareto-optimal solutions in one run of the program. Fourth, they are likely to find global optimal solutions. In general, MHAs have become quite popular for solving MOO problems in engineering and science (e.g., Radhika and Chaparala, 2018).

Like any method, MHAs also have several limitations. First, they use random numbers and so the final solutions can differ slightly from one run to another. Hence, it is desirable to solve the MOO problem by an MHA several times to find global optimal solutions. Second, although they are likely to approach global optimal solutions, they may not give precise solutions. Third, MHAs can handle inequality constraints satisfactorily, but they are not effective for equality constraints. Fourth, computational time for solving an MOO problem by MHAs may be more compared to the WSM and ε-constraint method in conjunction with efficient deterministic methods such as SQP and GRG, particularly for problems with two objectives. Fifth, while solving large-sized MOO application problems (e.g.,

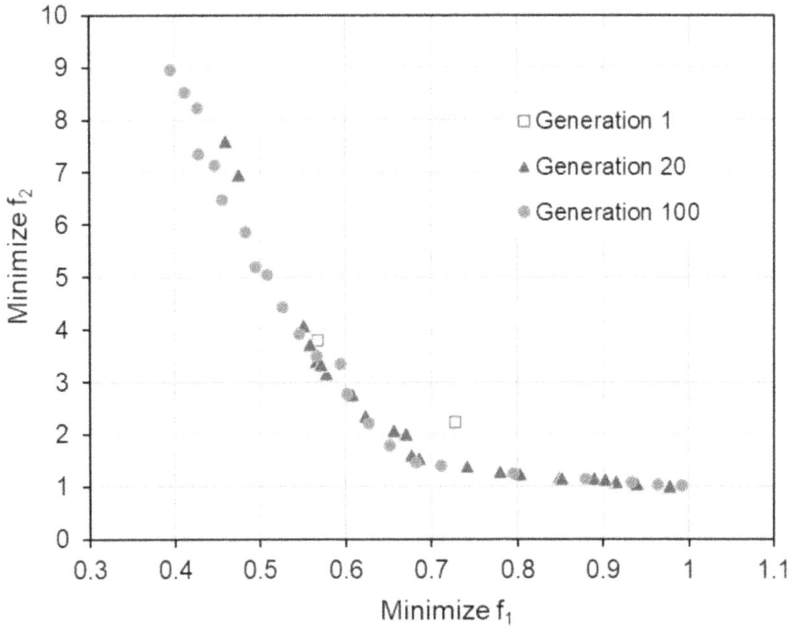

FIGURE 2.5
Solutions found by EMOO program for the CONSTR problem: Square symbols are after generation 1, triangle symbols are after 20 generations and circle symbols are after 100 generations.

more than ten objectives and more than 30 decision variables), MHAs can become (very) slow.

Example 2.3: Apply an MHA to solve the CONSTR problem in Eq. 2.2.

Solution: For solving the CONSTR problem, we choose to use the elitist NSGA-II of Deb (2001). Specifically, this MOO problem can be solved using the EMOO program, which is the NSGA-II code implemented in Microsoft Excel and described in the next section. Solutions found at different numbers of generations by this program using a population size of 20 are displayed in Figure 2.5.

Comments on results in Figure 2.5: Feasible solutions after generation 1 (square symbols) are only two and one of them is close to the solutions after 100 generations (circle symbols). One reason for this is that the CONSTR problem is relatively simple. Many MOO problems are more difficult and solutions after generation 1 may be fewer and/or may not be close to the converged (Pareto-optimal) solutions. For the CONSTR problem, after 20 generations (triangle symbols), there are many more solutions covering a reasonable range (Figure 2.5).

Finally, solutions after 100 are many and well-spread out over the widest range. Thus, the trade-off solutions improve and increase with generations.

2.5 Microsoft Excel, Python and MATLAB® Programs for Multi-Objective Optimization

Many programs are now readily available for MOO. With the ongoing developments in MOO, these may be updated and/or new programs will be added in the coming years. Here, a few selected programs for MOO in Microsoft Excel, Python and MATLAB platforms are outlined in Subsections 2.5.1, 2.5.2 and 2.5.3, respectively. See Section 5.3 in Rangaiah et al. (2020) for some MOO programs in other programming languages and/or platforms.

2.5.1 EMOO Program in Microsoft Excel

The Microsoft Excel program for MOO (EMOO), based on NSGA-II of Deb (2001) and binary coding of decision variables, was first developed and tested by Sharma et al. (2012). Then, it was enhanced by Wong et al. (2016) by the addition of real-coded NSGA-II and two termination criteria based on search progress. The EMOO program has been used extensively by the research students of the second author (GPR) of this book. Moreover, it is available freely and has been sent to several hundred engineers and scientists upon their request. Readers interested in EMOO program can contact Dr. Sharma at shivomsharma.iitr@gmail.com.

Main features of the EMOO program are outlined below. First, the program has a simple user interface and it can be used by anyone familiar with Microsoft Excel. Figure 2.6 shows the main interface for entering the data on objectives, decision variables and constraints as well as for changing the default values of algorithm parameters. The program has a separate worksheet for entering equations for objectives and constraints; this worksheet (Figure 2.7) shows the setup for the CONSTR problem (in the top half) and brief instructions (in the bottom half).

Second, the EMOO program has both binary and real-coded NSGA-II; either of them can be chosen for solving an MOO problem. Third, there is no limit on the number of objectives, decision variables and constraints in the problem, but the difficulty of solving the problem (i.e., computational time) increases. Fourth, in the EMOO program, inequality constraints are handled by the constrained-dominance principle, outlined in the previous section. If equality constraints are present in the problem, then users must incorporate them, for example, via a penalty function approach (Ramteke et al., 2021). Fifth, the program has two progress-based criteria besides the maximum

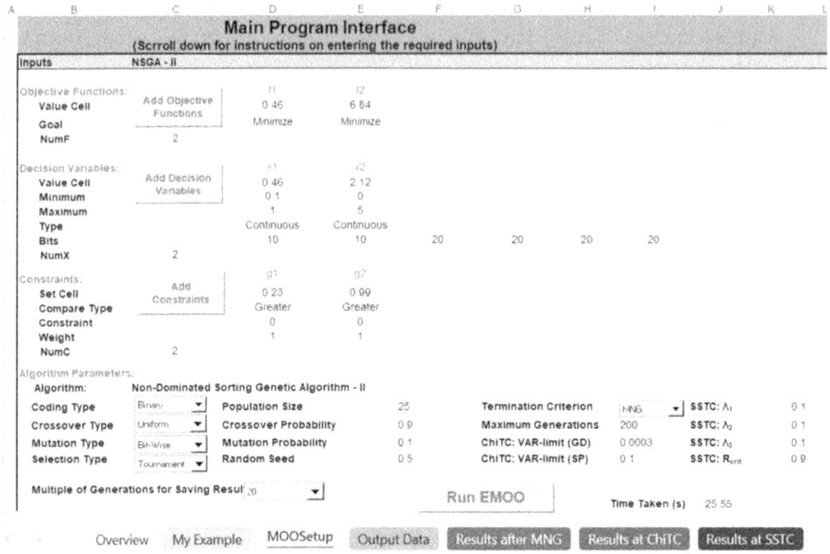

FIGURE 2.6
Main interface of the EMOO program for users to enter the data on objectives, decision variables and constraints; algorithm parameters are in the bottom part.

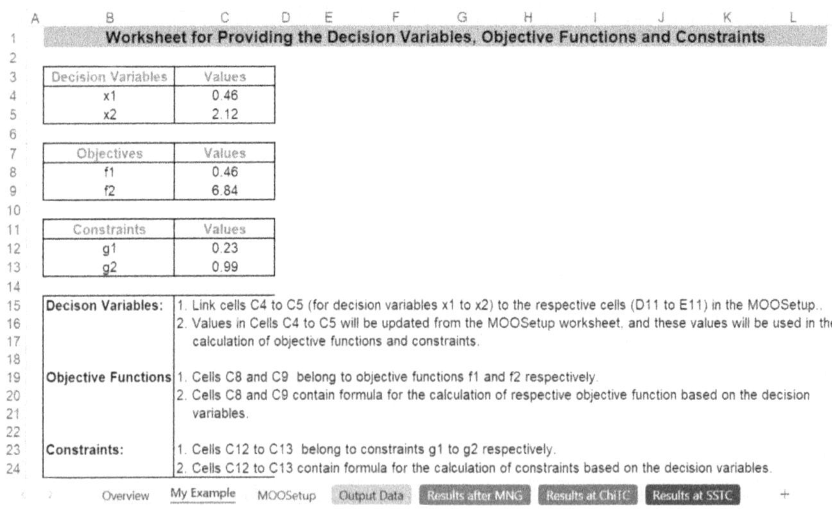

FIGURE 2.7
Worksheet in the EMOO program, for entering the decision variables and the equations for objectives and constraints.

	A	B	C	D	E	F	G	H	I	J	K
1		Note: Only feasible solutions from best/first Pareto-optimal front in multiple generations are presented.									
3		x1	x2	f1	f2	g1	g2				
4		***********Generation 1******************									
5		0.567155	1.15347	0.567155	3.796967	0.257869	2.950929				
6		0.728152	0.620723	0.728152	2.225802	1.174096	4.932649				
7											
8		***********Generation 20******************									
9		0.970968	0	0.970968	1.0299	2.73871	7.73871				
10		0.397361	2.502444	0.397361	8.814268	0.07869	0.073803				
11		0.488856	1.964809	0.488856	6.064787	0.364516	1.434897				
12		0.55044	1.251222	0.55044	4.08986	0.205181	2.702737				
13		0.460704	2.502444	0.460704	7.602376	0.648778	0.643891				
14		0.47478	2.30694	0.47478	6.965205	0.579961	0.96608				
15		0.742229	0.039101	0.742229	1.399974	0.719159	5.640958				
16		0.607625	0.674487	0.607625	2.755792	0.143109	3.794135				
17		0.62346	0.469208	0.62346	2.356538	0.080352	4.141935				
18		0.685924	0.068426	0.685924	1.557646	0.24174	5.104888				
19		0.577713	0.826002	0.577713	3.160745	0.025415	3.373412				
20		0.656891	0.371457	0.656891	2.087798	0.28348	4.540567				
21		0.558358	1.085044	0.558358	3.734244	0.110264	2.940176				
22		0.670968	0.356794	0.670968	2.022145	0.395503	4.681916				
23		0.84868	0	0.84868	1.1783	1.638123	6.638123				
24		0.88915	0.043988	0.88915	1.174142	2.046334	6.958358				
25		0.802933	0	0.802933	1.245435	1.226393	6.226393				
26		0.780059	0.014663	0.780059	1.300752	1.035191	6.005865				

Overview | My Example | MOOSetup | Output Data | Results after MNG | Results at ChiTC | Results at SSTC

FIGURE 2.8
Decision variables (x1, x2), objectives (f1, f2) and constraints (c1, c2) of the non-dominated solutions at different generations for the CONSTR problem.

number of generations for termination. Sixth, it presents results/solutions at intermediate generations and at each termination criterion in separate worksheets. Users can then prepare suitable plots (such as Figure 2.5) using the data on decision variables, objectives and constraints in these worksheets (e.g., Figure 2.8 for the CONSTR problem). Finally, a clear and concise user guide for the EMOO program is available.

Besides EMOO, another program, namely, the IMODE program in Microsoft Excel, is also available. It is based on DE enhanced with a tabu list and has self-adaptation of algorithm parameters, performance metrics and progress-based termination criteria, besides the maximum number of generations. These details along with the algorithm and flowchart are available in Sharma and Rangaiah (2013) and Sharma et al. (2017). The IMODE program is also available freely; interested readers can contact Dr. Sharma at shivomsharma.iitr@gmail.com.

2.5.2 Python Programs for MOO

The pymoo library, developed by Blank and Deb (2020), is a powerful and flexible Python library for solving MOO problems using a variety of MHAs. Designed for researchers and practitioners, it provides an intuitive and customizable environment for defining, solving and analyzing MOO problems.

It supports a wide range of algorithms, including NSGA-II, NSGA-III and various other state-of-the-art evolutionary approaches. The library is built on top of NumPy with efficient computations. It includes modules for problem definition, solution evaluation, visualization and performance benchmarking. Its modular architecture allows users to implement custom operators, decision variables, constraints, optimization objectives and termination criteria, making it highly adaptable to problem domains in engineering and science.

One of the special features of pymoo is its support for both SOO and MOO scenarios. It offers built-in visualization tools that allow users to analyze Pareto-optimal fronts, convergence behavior and performance metrics. Additionally, pymoo includes a benchmarking module for comparing different algorithms across various test problems. The Python code in Table 2.6 demonstrates the formulation and solution of the CONSTR problem using the NSGA-II method implemented in the pymoo library. Its executed results after 1, 20 and 100 generations are presented in Figure 2.9. After generation 1, there are only five feasible solutions, which are much inferior to non-dominated solutions after generations 20 and 100. Results after 20 generations are quite good; they are slightly inferior to non-dominated solutions after 100 generations in terms of both objective values and distribution of solutions.

2.5.3 Multi-Objective Optimization Programs in MATLAB

The commercial software MATLAB has many capabilities and applications for the simulation and analysis of engineering and scientific problems. It has several programs for solving (constrained) MOO problems. These are summarized in Table 2.7.

2.6 Comparison Between Multi-Objective Optimization and Multi-Criteria Decision-Making

Before concluding this chapter, the MCDM introduced in Chapter 1 and the MOO described in this chapter are compared in Table 2.8 in terms of many factors such as purpose, problem aspects, solution methods and their nature, results obtained and the role of decision-makers. Whereas MCDM focuses on ranking and selecting from a set of alternatives, taking multiple criteria into consideration, MOO aims to search for and identify a set of equally good Pareto-optimal solutions by optimizing multiple conflicting objectives simultaneously (Wang et al., 2023). Both MCDM and MOO fall under optimization, going by the dictionary meaning of optimization: finding the best or most effective solution.

TABLE 2.6

Python Code Using the NSGA-II Method in the pymoo Library (Version 0.6.1.3).

```
import numpy as np
import pandas as pd
from pymoo.algorithms.moo.nsga2 import NSGA2
from pymoo.core.problem import ElementwiseProblem
from pymoo.optimize import minimize
from pymoo.visualization.scatter import Scatter

class CONSTR(ElementwiseProblem):

    def __init__(self):
        super().__init__(n_var=2,
                         n_obj=2,
                         n_ieq_constr=2,
                         xl=np.array([0.1, 0.0]), # lower bound
                         xu=np.array([1.0, 5.0])) # upper bound

    def _evaluate(self, x, out, *args, **kwargs):
        #  In pymoo, each objective function is supposed to be minimized
        f1 = x[0]
        f2 = (1 + x[1]) / x[0]

        # each constraint needs to be provided in the form of <= 0
        g1 = -9 * x[0] - x[1] + 6.0
        g2 = -9 * x[0] + x[1] + 1.0

        out["F"] = [f1, f2]
        out["G"] = [g1, g2]

problem = CONSTR()

algorithm = NSGA2(pop_size=25)

res = minimize(problem,
               algorithm,
               ("n_gen", 100), # 100 generations
               verbose=False,
               seed=1)

sols = np.hstack((res.X, res.F))
sorted_sols = sols[np.argsort(sols[:, 2])] # sort by f1
pd.options.display.float_format = '{:.6f}'.format # show 6 decimal places
df_sols = pd.DataFrame(sorted_sols, columns=["x1", "x2", "f1", "f2"])
display(df_sols)

plot = Scatter()
plot.add(res.F, edgecolor="black", facecolor="none")
plot.show()
```

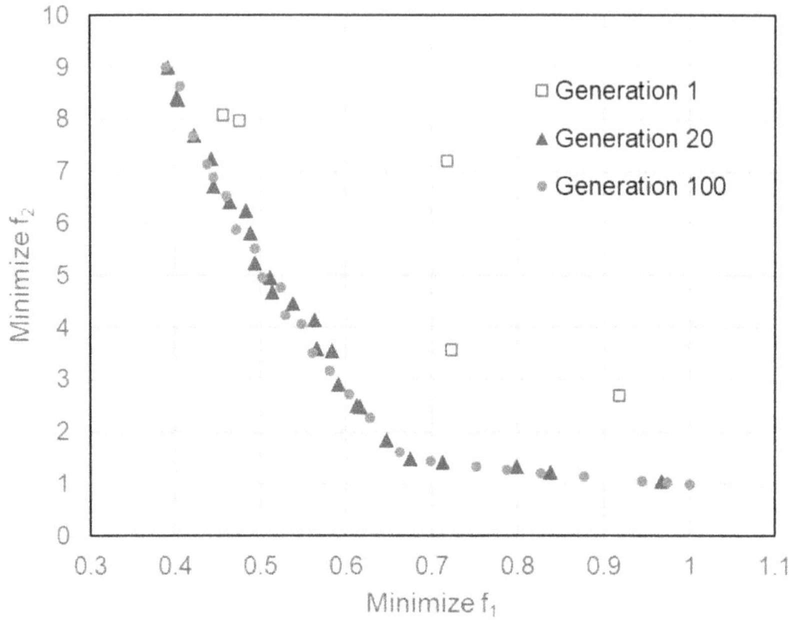

FIGURE 2.9
Results obtained using the Python code for the CONSTR problem.

TABLE 2.7
MATLAB Programs for Solving MOO Problems.

Name	Brief Details (with Reference)
GAMULTIOBJ	This program in the global optimization toolbox of MATLAB employs NAGA-II for solving MOO problems and presents real-time plots to monitor the progress of optimization. (https://au.mathworks.com/help/gads/gamultiobj.html, accessed in February 2025).
MOSQP	It is based on SQP for solving differentiable, constrained MOO problems. So, the method in this program belongs to a single-objective approach (MOSQP Solver available at MOSQP Home Page, accessed in February 2025).
NGPM	Reference-point-based NSGA-II Program effective for solving MOO problems (https://au.mathworks.com/matlabcentral/fileexchange/31166-ngpm-a-nsga-ii-program-in-matlab-v1-4, accessed in February 2025).
NSGA-III	Implementation of NSGA-III that is more effective (than NSGA-II) for MOO problems having three or more objectives (https://yarpiz.com/456/ypea126-nsga3, accessed in February 2025).
PlatEMO	This extensive software for solving MOO problems has more than 50 multi-objective evolutionary algorithms, more than 100 MOO test problems and several popular performance indicators (Tian et al., 2017).
TOMLAB	An optimization environment in MATLAB supporting various problem types, including MOO solvers like genetic and evolutionary optimization algorithms (https://tomopt.com/tomlab/, accessed in February 2025).

TABLE 2.8

Comparison of MCDM and MOO.

Factor	MCDM	MOO
Purpose	Sort, rank and/or select one from the given set of alternatives whose performance is quantified in terms of several criteria.	Find the non-dominated (Pareto-optimal) solutions as well as the values of decision variables and constraints for optimizing the objectives of a given application.
Problem aspects	The problem involves only alternatives and criteria (e.g., see example applications in Appendix A). In MOO, the equivalent of an alternative is a Pareto-optimal solution and the criteria are generally termed objectives.	The problem involves objectives, continuous and/or discrete variables, bounds on them and equality and inequality constraints (Eq. 2.1). Allowable values of decision variables are limited by bounds and constraints. They are critical in achieving desired objective values.
Problem size	Commonly used in contexts with a moderate or small number of alternatives (dozens or hundreds). Problems with thousands of alternatives can still be handled but are not commonly seen in practice.	Can handle high-dimensional decision spaces (many variables, large ranges, various constraints and many Pareto-optimal solutions obtained), although computationally intensive.
Methods for solving the problem	Several hundred MCDM methods (Baydas et al., 2024) for sorting and ranking the alternatives; some of them (e.g., GP) can be used to find the most preferred non-dominated solution.	Many methods (e.g., WSM, ε-constraint method and MHAs) for solving the MOO problem. A few others (e.g., GP) can be interpreted as an MOO method, since no alternative needs to be provided a priori.
Nature of methods	These methods are often neither iterative nor computationally intensive.	These methods are generally iterative and computationally intensive.
Results or Output	Ranks of alternatives that can be utilized for selection and/or sorting.	Non-dominated (or Pareto-optimal) solutions (i.e., values of objectives as well as values of decision variables to achieve them), representing the trade-off between objectives.
Role of decision-makers	Typically requires the decision-makers to express preferences during the ranking process (e.g., criteria weights).	Decision-makers are to visualize and analyze Pareto-optimal solutions after solving the MOO problem.
Interpretation of results	Provides an easily interpretable ranking (e.g., alternative A is top-ranked and B is second-ranked).	Outputs a set of equally good non-dominated solutions. It can be somewhat challenging for nontechnical stakeholders to interpret.

MCDM is often used in decision-making scenarios, requiring explicit information like weights from decision-makers (Wang et al., 2024), whereas MOO handles constrained, high-dimensional spaces, which typically need iterative and computationally intensive methods. Nonetheless, MOO and MCDM can be effectively integrated in practice (e.g., Nabavi et al., 2023), with MOO generating the Pareto-optimal front and MCDM ranking and selecting one or several solutions for ultimate implementation.

After finding many Pareto-optimal (non-dominated) solutions for an MOO application, MCDM is necessary to choose the most preferred optimal solution. Therefore, one may question the need to find many/all Pareto-optimal solutions. Indeed, it is possible to find the most preferred optimal solution using methods such as GP (covered in the next chapter), WSM and ε-constraint method, without having many Pareto-optimal solutions. However, decision-makers must provide the right weights (in the WSM), ε values (in the ε-constraint method) and acceptable goals for objectives to be achieved (in the GP). Providing any of these, without some knowledge of Pareto-optimal solutions, is challenging to decision-makers. Hence, MOO is beneficial in applications where alternatives are not available, for example, through surveys of the market and/or literature.

Indeed, there are several benefits of having many Pareto-optimal solutions by solving the MOO problem for an application. First, they provide deeper knowledge on the optimal values of objectives, their variations and ranges, which could be useful for providing weights and/or other inputs in the MCDM procedure described in Chapter 1. Second, any MCDM method can be applied to Pareto-optimal solutions. Then, the recommended solutions by several MCDM methods can be reviewed collectively for choosing the most preferred optimal solution. Third, sensitivity analysis in the MCDM procedure can be conducted easily. Fourth, having the entire Pareto-optimal front can reveal interesting patterns and insights about trade-offs (e.g., "objective A can only be improved by sacrificing a small amount in objective B, but objective C deteriorates a lot"). Such insights might be overlooked if only a single solution were sought.

2.7 Summary

Key concepts and methods described in this chapter are summarized below.

1. Both SOO and MOO problems share the common features of continuous/discrete decision variables, bounds on variables and equality/inequality constraints. The main difference between them is in the number of objectives; this and other (consequential) differences are summarized.
2. Two or more objectives in MOO and the presence of conflict among the objectives lead to non-dominated (Pareto-optimal) solutions, having trade-offs among the objectives.

3. MOO problems can be solved using the methods for SOO, for example, through WS and ε-constraint methods. These methods belong to the single-objective approach, find only one Pareto-optimal solution from each run and are satisfactory for MOO problems with two or three objectives.
4. MHAs have been successfully adopted for solving MOO problems; they yield many non-dominated solutions in one run of the program. These belong to the multi-objective approach and are effective even for MOO problems with a large number of objectives.
5. A Microsoft Excel program based on the NSGA-II and the pymoo library in Python, for solving MOO problems, are described. Several MATLAB programs for solving MOO problems are listed.
6. Finally, differences between MCDM and MOO in terms of many factors, such as purpose, problem aspects, methods for solving them, results obtained and the role of decision-makers, are tabulated. The benefits of finding many non-dominated solutions are also outlined.

References

Baydas, M., Kavacık, M., & Wang, Z. (2024). Does the performance of MCDM rankings increase as sensitivity decreases? Graphics card selection and pattern discovery using the PROBID method. *Journal of Intelligent Management Decision, 3*(2), 91–103. https://doi.org/10.56578/jimd030203

Blank, J., & Deb, K. (2020). Pymoo: Multi-objective optimization in python. *IEEE Access, 8*, 89497–89509.

Deb, K. (2001). *Multi-objective optimization using evolutionary algorithms.* Chichester, UK: John Wiley & Sons.

Deb, K., Pratap, A., Agarwal, A., & Meyarivan, T. (2002). A fast and elitist multi-objective genetic algorithm: NSGA-II. *IEEE Transactions on Evolutionary Computation, 6*, 182–197.

Kim, I. Y., & de Weck, O. L. (2005). Adaptive weighted-sum method for bi-objective optimization: Pareto front generation. *Structural and Multidisciplinary Optimization, 29*, 149–158.

Miettinen, K. (1999). *Nonlinear multi-objective optimization.* Boston: Kluwer Academic Publishers.

Nabavi, S. R., Jafari, M. J., & Wang, Z. (2023). Deep learning aided multi-objective optimization and multi-criteria decision making in thermal cracking process for olefines production. *Journal of the Taiwan Institute of Chemical Engineers, 152*, 105179.

Nabavi, S. R., Wang, Z., & Rodríguez, M. L. (2024). Multi-objective optimization and multi-criteria decision-making approach to design a multi-tubular packed-bed membrane reactor in oxidative dehydrogenation of ethane. *Energy & Fuels, 39*(1), 491–503.

Park, M., Wang, Z., Li, L., & Wang, X. (2023). Multi-objective building energy system optimization considering EV infrastructure. *Applied Energy, 332*, 120504.

Radhika, S., & Chaparala, A. (2018). Optimization using evolutionary metaheuristic techniques: A brief review. *Brazilian Journal of Operations & Production Management, 15*(1), 44–53. https://doi.org/10.14488/BJOPM.2018.v15.n1.a17

Rajwar, K., Deep, K., & Das, S. (2023). An exhaustive review of the metaheuristic algorithms for search and optimization: Taxonomy, applications and open challenges. *Artificial Intelligence Review, 56*, 13187–13257. https://doi.org/10.1007/s10462-023-10470-y

Ramteke, M., Saraf, D. N., & Gupta, S. K. (2021). *Optimization for engineers*. London: New Age International Publishers.

Rangaiah, G. P. (Ed.). (2017). *Multi-objective optimization: Techniques and applications in chemical engineering* (2nd ed., Vol. 5 in the Advances in Process Systems Engineering). World Scientific.

Rangaiah, G. P., Feng, Z., & Hoadley, A. F. (2020). Multi-objective optimization applications in chemical process engineering: Tutorial and review. *Processes, 8*, 508. https://doi.org/10.3390/pr8050508

Rao, S. S. (2009). *Engineering optimization: Theory and practice* (4th ed.). John Wiley.

Sharma, S., & Rangaiah, G. P. (2013). An improved multi-objective differential evolution with a termination criterion for optimizing chemical processes. *Computers and Chemical Engineering, 56*, 155–173.

Sharma, S., Rangaiah, G. P., & Cheah, K. S. (2012). Multi-objective optimization using Microsoft Excel with an application to design of a falling-film evaporator system. *Food and Bioproducts Processing, 90*, 123–134.

Sharma, S., Rangaiah, G. P., & Maréchal, F. (2017). Integrated multi-objective differential evolution and its application to amine absorption process for natural has sweetening. In G. P. Rangaiah, & S. Sharma (Eds.), *Differential evolution in chemical engineering: Developments and applications*. World Scientific.

Tian, Y., Cheng, R., Zhang, X., & Jin, Y. (2017). PlatEMO: A MATLAB® platform for evolutionary multi-objective optimization. *IEEE Computational Intelligence Magazine, 12*, 73–87.

Virtanen, P., Gommers, R., Oliphant, T. E., Haberland, M., Reddy, T., Cournapeau, D., … & Van Mulbregt, P. (2020). SciPy 1.0: Fundamental algorithms for scientific computing in Python. *Nature Methods, 17*(3), 261–272.

Wang, Y., Wang, Z., & Wu, Z. (2025). Multi-objective optimal control of nonlinear processes using reinforcement learning with adaptive weighting. *Computers & Chemical Engineering*, 109206.

Wang, Z., Li, J., Rangaiah, G. P., & Wu, Z. (2022). Machine learning aided multi-objective optimization and multi-criteria decision making: Framework and two applications in chemical engineering. *Computers & Chemical Engineering, 165*, 107945.

Wang, Z., Nabavi, S. R., & Rangaiah, G. P. (2024). Multi-criteria decision making in chemical and process engineering: Methods, progress and potential. *Processes, 12*(11), 2532.

Wang, Z., Parhi, S. S., Rangaiah, G. P., & Jana, A. K. (2020). Analysis of weighting and selection methods for pareto-optimal solutions of multiobjective optimization in chemical engineering applications. *Industrial & Engineering Chemistry Research, 59*(33), 14850–14867.

Wang, Z., Tan, W. G. Y., Rangaiah, G. P., & Wu, Z. (2023). Machine learning aided model predictive control with multi-objective optimization and multi-criteria decision making. *Computers & Chemical Engineering, 179*, 108414.

Wong, J. Y., Sharma, S., & Rangaiah, G. P. (2016). Design of shell-and-tube heat exchangers for multiple objectives using elitist non-dominated sorting genetic algorithm with termination criteria. *Applied Thermal Engineering, 93*, 888–899.

Exercises

E2.1 Apply the WSM with w_1 = 0.1, 0.2, 0.4, 0.6, 0.8 and 0.9 to solve the CONSTR problem. Comment on the Pareto-optimal solutions found using these weights and then compare them with those in Table 2.1.

E2.2 Reproduce the results given in Example 2.2 by the ε-constraint method and Example 2.3 by MHA.

E2.4 Apply the (1) WSM, (2) ε-constraint method and (3) MHA to solve the Osyckza problem given below. This problem is difficult to solve compared to the CONSTR problem.

Maximize $$f_1(\mathbf{x}) = \left[25(x_1-2)^2 + (x_2-2)^2 + (x_3-1)^2 + (x_4-4)^2 + (x_5-1)^2\right]$$

Minimize $$f_2(\mathbf{x}) = x_1^2 + x_2^2 + x_3^2 + x_4^2 + x_5^2 + x_6^2$$

With respect to: $(x_1,\, x_2,\, x_6) \in [0,10],\ (x_3, x_5) \in [1,5],\ x_4 \in [0,6]$

Subject to:

$$g_1(\mathbf{x}) = x_1 + x_2 - 2 \geq 0$$

$$g_2(\mathbf{x}) = -x_1 - x_2 + 6 \geq 0$$

$$g_3(\mathbf{x}) = x_1 - x_2 + 2 \geq 0$$

$$g_4(\mathbf{x}) = -x_1 + 3x_2 + 2 \geq 0$$

$$g_5(\mathbf{x}) = -(x_3-3)^2 - x_4 + 4 \geq 0$$

$$g_6(\mathbf{x}) = (x_5-3)^2 + x_6 - 4 \geq 0$$

In the above problem, decision variables and their bounds are given together and there are no equality constraints.

E2.5 Apply the (1) WSM, (2) ε-constraint method and (3) MHA to solve the following the WATER problem with three decision variables, five objectives and seven inequality constraints (but no equality constraints). This problem with five objectives is tedious to solve by the WSM and ε-constraint method (compared to MHA).

Minimize $$f_1(\mathbf{x}) = 106780.37(x_2 + x_3) + 61704.67$$

Minimize $$f_2(\mathbf{x}) = 3000x_1$$

Minimize $f_3(\mathbf{x}) = (305700)2289x_2 / [(0.06)2289]^{0.65}$

Minimize $f_4(\mathbf{x}) = (250)2289\exp(-39.75x_2 + 9.9x_3 + 2.74)$

Minimize $f_5(\mathbf{x}) = 25(1.39/(x_1x_2) + 4940x_3 - 80)$

With respect to:
$$0.01 \le x_1 \le 0.45$$
$$0.01 \le x_2 \le 0.1$$
$$0.01 \le x_3 \le 0.1$$

Subject to:

$$g_1(\mathbf{x}) = 0.00139/(x_1x_2) + 4.94x_3 - 0.08 \le 1.00$$

$$g_2(\mathbf{x}) = 0.000306/(x_1x_2) + 1.082x_3 - 0.0986 \le 1.00$$

$$g_3(\mathbf{x}) = 12.307/(x_1x_2) + 49408.24x_3 + 4051.02 \le 50000.00$$

$$g_4(\mathbf{x}) = 2.098/(x_1x_2) + 8046.33x_3 - 696.71 \le 16000.00$$

$$g_5(\mathbf{x}) = 2.138/(x_1x_2) + 7883.39x_3 - 705.04 \le 10000.00$$

$$g_6(\mathbf{x}) = 0.417/(x_1x_2) + 1721.26x_3 - 136.54 \le 2000.00$$

$$g_7(\mathbf{x}) = 0.164/(x_1x_2) + 631.13x_3 - 54.48 \le 550.00$$

E2.6 Outline the merits and limitations of single- and multi-objective approaches for solving MOO problems. If necessary, see Section 5.1.4 in Rangaiah et al. (2020).

E2.7 Describe the similarities and differences between the WSM and the ε-constraint method.

E2.8 Which method between the WSM and the ε-constraint method will you recommend and why?

E2.9 Which MOO method will you recommend for solving each of the following MOO problems? Justify your answer for each MOO problem.

A. MOO problem with two objectives; decision variables, objectives and constraints are all continuous.
B. MOO problem with two objectives; decision variables are continuous and derivatives of objectives and constraints are not available.
C. MOO problem with five objectives; decision variables, objectives and constraints are all continuous.
D. MOO problem with five objectives; decision variables are continuous and derivatives of objectives and constraints are not available.

3

Goal Programming

3.1 Overview

Goal programming (GP) is one of the popular methods in the field of multi-criteria decision-making (MCDM). It is typically used for solving multi-objective linear optimization problems (i.e., every objective function and every constraint is a linear expression of decision variables) using a linear programming (LP) method such as the Simplex algorithm. *A linear expression is one in which each variable appears with a constant coefficient and no variable is multiplied by another variable, raised to a power, placed in a denominator, or wrapped inside a nonlinear function such as an absolute value, square root, exponential, or trigonometric term.* Hence, GP can be considered as an LP method for multiple objectives, as a method for multi-objective optimization (MOO), as well as a method in the multi-objective decision-making (MODM) group of MCDM (Section 1.1). The approach of GP can be extended to solving nonlinear MOO problems.

Although the GP method was first employed in Charnes et al. (1955), it was termed and described as GP later in the book by Charnes and Cooper (1961). There are now many variants of GP; of these, the basic ones are weighted, lexicographic and min–max (also known as Chebyshev) GPs. Numerous applications of GP in diverse fields have been reported. For example, many applications in accounting, agriculture, economics, engineering, finance, government, international context, management and marketing are listed in Chapter 4 of the book by Schniederjans (1994). More recently, Colapinto et al. (2015) reviewed about 300 applications of GP variants in engineering, management and social sciences.

The main aims of this chapter are to (1) present GP concepts and models, (2) describe three basic variants of GP with an example, (3) illustrate GP for applications wherein ACM data are available and (4) discuss the merits and limitations of GP as well as guidelines for the successful application of GP. The presentation of these is limited to linear GP. Accordingly, sections of this chapter are:

Section 3.2 Goal Programming Concepts and Problem
Section 3.3 Weighted Goal Programming
Section 3.4 Lexicographic Goal Programming
Section 3.5 Min–Max (Chebyshev) Goal Programming

DOI: 10.1201/9781003530435-3

The learning outcomes of this chapter on GP are:

1. Describe GP and its models.
2. Formulate a GP model for an application and solve it using Microsoft Excel or a Python program.
3. Discuss the merits and limitations of GP.
4. Compare GP with related methods.

The presentation in this chapter assumes that readers are familiar with LP, its applications, methods and the solution of LP problems using, for example, the Simplex LP option available in the Solver tool in Microsoft Excel. If required to learn these, identify and study any good textbook on optimization. For example, Chapter 7 in Edgar et al. (2001) is on LP and it includes an example using the Solver in Microsoft Excel.

Notation and terminology in the present chapter are mostly the same as in the rest of this book, but there are slight differences (e.g., criteria and objectives are synonymous in other chapters, but, in this chapter, the objective function is a function of deviation variables that are specifically introduced in GP). For more details on GP and its applications, the comprehensive and concise book by Jones and Tamiz (2010) is suggested, particularly for readers new to GP. It covers many variants of GP, formulation of GP problems, solution of GP problems using Excel's Solver and LINGO, detection and restoration of Pareto inefficient (i.e., dominated) solutions found by GP, integration of GP with AI, data envelopment analysis, etc. and two detailed case studies (one in health care and another in portfolio selection) in separate chapters.

3.2 Goal Programming Concepts and Problem

A typical MCDM problem, introduced in Chapter 1, has m alternatives and n criteria, each of which should be maximized or minimized (i.e., optimized). *In the GP covered in the present chapter, a goal to be achieved for each criterion is introduced.* This goal is usually lower than the maximum possible value (in case of a maximization criterion) or more than the minimum possible value (in case of a minimization criterion). Thus, optimization of criteria is changed to satisficing (i.e., achieving satisfactory values for the criteria, as given by the corresponding goals). For example, the profit criterion that should be maximized (say, with a maximum possible value of \$550,000/month) is changed to a profit goal with a certain value (say, \$480,000/month).

Depending on the situation, the specified goal value may be exceeded or may not be achieved. To allow for these, *positive and negative deviation variables (P and N) are introduced in the equation for each goal.* These deviation variables are restricted to non-negative values (i.e., ≥ 0). For example, the equation for profit goal is modified to introduce both a negative deviation (that corresponds to failing to achieve the given goal value of \$480,000) and a positive deviation (that corresponds to exceeding the given goal value). Obviously, positive deviation is desirable, whereas negative deviation should be avoided. In some other cases (e.g., manufacturing time or energy), positive deviation (i.e., exceeding the goal value) is not desirable and should be avoided, whereas negative deviation (i.e., keeping within the goal value) is acceptable.

The equation for each goal with a positive deviation variable (P) and a negative deviation variable (N) is termed a *soft constraint* in GP. So, there will be n soft constraints in a GP problem with n criteria or goals. These constraints are of equality type. Further, a GP problem will have *decision variables* (denoted by a vector, **x**) that can be continuous and/or discrete; for simplicity, separate symbols are <u>not</u> employed for them in this chapter. Values of these decision variables determine the goal value through the respective equations (i.e., equations for goals are functions of **x** besides **P** and **N**). In GP, decision variables are different from and in addition to the deviation variables introduced in the previous paragraph. However, like deviation variables, decision variables are also restricted to non-negative values (i.e., ≥ 0).

For example, suppose a goal of 850 is given by the summation of two decision variables as follows: $x_1 + x_2 = 850$. The introduction of deviation, "d", changes this equation to $x_1 + x_2 - 850 = d$, which can be rewritten as $x_1 + x_2 - d = 850$. Here, d can be negative, zero, or positive and so it can be expressed as the difference of two non-negative values (i.e., $d = P - N$). Hence, the goal equation becomes $x_1 + x_2 - (P - N) = 850$ or $x_1 + x_2 + N - P = 850$. Eq. 3.1d below is the generalization of this goal equation.

Often, GP applications have some constraints that must be satisfied. These constraints are functions of decision variables (and not deviation variables) and can be of either equality or inequality type. They are referred to as *hard (or system) constraints*. Finally, a GP problem has an *objective function* of deviation variables (Ns and Ps) only; this objective function is entirely different from the criteria.

In summary, the general form of a GP problem is as follows:

$$\text{Minimize} \quad f(\mathbf{N}, \mathbf{P}) \tag{3.1a}$$

$$\text{With respect to} \quad \mathbf{N}, \mathbf{P}, \mathbf{x} \tag{3.1b}$$

Subject to $\mathbf{N} \geq 0, \mathbf{P} \geq 0, \mathbf{x} \geq 0$ (3.1c)

$$C_i(\mathbf{x}) + N_i - P_i = G_i \quad \text{for } i = 1, 2, 3, \ldots, n \qquad (3.1d)$$

Hard / System constraints (3.1e)

In the above equations, n is the number of goals (also equal to the number of criteria); **N** and **P** are vectors of negative and positive deviation variables and they have n elements of N_i and P_i, respectively; f(**N**, **P**) is the objective function involving **N** and **P** only; $C_i(\mathbf{x})$ is a function of decision variables (**x**) and G_i is the goal, both for the ith criterion. The n equations (Eq. 3.1d) with one equation for each goal are soft constraints. Clearly, GP essentially formulates the optimization problem into one objective function, multiple decision variables, many deviation variables and multiple constraints.

The objective function (Eq. 3.1a), all $C_i(\mathbf{x})$ in Eq. 3.1d and hard/system constraints (Eq. 3.1e) are often assumed to be linear functions of decision variables (**x**) and deviation variables (**N** and **P**). In other words, the GP problem is presented as an LP problem. However, all or some $C_i(\mathbf{x})$ and hard/system constraints can be nonlinear functions of **x** (i.e., nonlinear GP). In this chapter, only linear GP problems and selected methods for their solution are described and illustrated in the subsequent sections; however, for brevity, the adjective, linear before GP, is not added. The methods described in this chapter require the solution of one or more LP problems, say, by the Simplex method. Hence, GP methods can be considered as the methods for solving LP problems having multiple criteria to be optimized.

Example 3.1: Consider a typical application for reinforcing the above concepts of GP. A semiconductor manufacturing plant produces three types of memory chips: Basic, Advanced and Premium. The time required for key steps in manufacturing one chip of these types as well as the profit per chip is given in Table 3.1.

Based on equipment and manpower resources, the plant currently has 800 hours of photolithography time and 1200 hours of testing time available daily. Of these, additional testing time is possible using spare equipment and

TABLE 3.1

Types of Chips, Required Photolithography and Testing Time and Profit.

Chip Type	Photolithography Time (h/chip)	Testing Time (h/chip)	Profit ($/chip)
Basic	0.8	1.7	54
Advanced	1.4	2.3	83
Premium	2.0	3.1	127

overtime by plant operators. Additional photolithography time is not possible due to the expensive equipment required for this step.

Considering the market demand, the plant management has set strategic goals, priorities and preference weights in Table 3.2; these are required for GP methods presented in the subsequent sections. A lower priority level (such as 1 and 2) corresponds to greater importance and so a larger preference weight. Thus, both priority level and preference weight are similar and are given as they are employed in some GP methods. Specifically, a priority level is required in lexicographic GP, whereas preference weights are needed in weighted GP.

For the above scenario, formulate a GP problem/model for finding the number of memory chips of each of the three types to be produced daily.

Solution: This example is set in the context of GP (e.g., goals are stated without stating the criteria, decision variables are hinted and alternatives are not given); note that alternatives are given in many MCDM application problems. Here, the formulation of the GP problem begins with the identification of decision variables, **x**, which will be helpful in developing the equations for soft and hard constraints. From the scenario description, the appropriate **x's** are:

x_B – Number of Basic chips to be produced per day

x_A – Number of Advanced chips to be produced per day

x_P – Number of Premium chips to be produced per day

Obviously, each of these three decision variables should be ≥ 0 as well as an integer.

Now, the equation for Goal 1 (i.e., total number of chips/day) can be developed as follows:

$$x_B + x_A + x_P + N_1 - P_1 - 600 \tag{3.2a}$$

In the optimum solution, either N_1 or P_1 will be zero. If $N_1 > 0$, then the total number of chips produced (i.e., $x_B + x_A + x_P$) will be < 600. On the other hand,

TABLE 3.2

Strategic Goals, Priority Levels and Preference Weights for Chip Manufacture.

Goal	Priority Level	Preference Weight
Goal 1 – Manufacture at least 600 chips/day	1	4
Goal 2 – Achieve a profit of at least $60,000/day	1	4
Goal 3 – Minimize additional testing time required	2	3
Goal 4 – Manufacture no more than 160 Premium chips per day	3	2

if $P_1 > 0$, then the total number of chips produced will be > 600 due to the negative coefficient of P_1 in the above equation. Since the goal is to manufacture at least 600 chips/day, producing fewer than the given goal (captured by the negative deviation, N_1) is undesirable and hence, N_1 should be minimized. On the other hand, producing more than 600 chips (captured by the positive deviation variable, P_1) is acceptable and so it is not penalized, meaning P_1 does not need to be minimized and can take any value ≥ 0.

Next, using the profit ($/chip) data in Table 3.1, the equation for Goal 2 (i.e., profit) is:

$$54x_B + 83x_A + 127x_P + N_2 - P_2 = 60,000 \tag{3.2b}$$

The significance of N_2 and P_2 is like that of N_1 and P_1. Further, considering the goal to achieve is a profit of $60,000/day, N_2 should be minimized, whereas P_2 can take any positive value.

Then, based on the testing time (h/chip) data in Table 3.1, the equation for Goal 3 (i.e., total testing time) is:

$$1.7x_B + 2.3x_A + 3.1x_P + N_3 - P_3 = 1200 \tag{3.2c}$$

Unlike Goals 1 and 2, Goal 3 is to minimize additional testing time. Hence, P_3, representing undesirable additional testing time beyond the allocated 1200 hours, should be minimized. On the other hand, N_3, indicating unused testing hours, is not penalized and can take any value ≥ 0.

Finally, the equation for Goal 4 (i.e., number of Premium chips manufactured) is:

$$x_P + N_4 - P_4 = 160 \tag{3.2d}$$

Since this goal is to manufacture no more than 160 Premium chips per day, N_4 can take any positive value, whereas P_4 should be minimized.

Besides the above soft constraints, will there be any system/hard constraints in this application? Careful reading of the scenario gives one hard constraint for not exceeding 800 hours of lithography time. This constraint can be expressed as follows:

$$0.8x_B + 1.4x_A + 2.0x_P \leq 800 \tag{3.2e}$$

Based on the above goal equations and discussion, the objective function will be a function of N_1, N_2, P_3 and P_4 (i.e., deviation variables to be minimized). This completes the formulation of the GP problem, which can be summarized as follows:

$$\text{Minimize} \quad f(N_1, N_2, P_3, P_4) \tag{3.3a}$$

$$\text{With respect to} \quad N_1, N_2, N_3, N_4, P_1, P_2, P_3, P_4, x_B, x_A, x_P \tag{3.3b}$$

$$\text{Subject to} \quad \begin{array}{l} N_1 \geq 0\, N_2 \geq 0, N_3 \geq 0, N_4 \geq 0, P_1 \geq 0, \\ P_2 \geq 0, P_3 \geq 0, P_4 \geq 0 \text{ and} \\ x_B, x_A, x_P \geq 0 \text{ and Integer} \end{array} \tag{3.3c}$$

$$\text{Eqs. 3.2a to 3.2e} \tag{3.3d}$$

In the above optimization problem, there are eight deviation variables and three decision variables that can be varied to minimize the objective function, having only four of the eight deviation variables. However, all deviation and decision variables occur in one or more soft and/or hard constraints. The objective function formulation will be described in the following sections.

Comments: The GP problem (i.e., equations or model) depends on the scenario. It is essential to understand the application correctly and include all relevant equations (constraints) and decision and deviation variables. This requires a good background in the domain of the application. Any error in the equations, variables and/or numerical values will lead to an optimal solution that will be incorrect and inapplicable. If details as in Example 3.1 are available, then the GP problem formulation is not difficult.

Often, the objective function in GP (Eq. 3.3a) is linear in the deviation variables and it can take several forms. As shown in Eq. 3.3a, either N_i or P_i (and not both) in the ith goal equation will be involved in the objective function. The GP problem formulation for another application on the manufacture of shirts and jackets is described in detail in Section 6.2.1 of the book by Papathanasiou and Ploskas (2018). Study this application for further learning of GP problem formulation. For GP examples in diverse areas, see the exercises at the end of Chapter 3 in the book by Jones and Tamiz (2010).

3.3 Weighted Goal Programming

There are many variants of linear GP that differ in the objective function (Eq. 3.1a) but have the same set of decision and deviation variables as well as constraints (i.e., Eqs. 3.1b–3.1e are the same). Among these variants are the distance-based variants of GP. Weighted GP, also known as non-preemptive GP, is one of the distance-based variants of GP. In this variant, the objective function (Eq. 3.1a) is taken to be the weighted sum (WS) of deviation variables to be minimized.

Instead of a simple sum of deviation variables, WS is required for two reasons. One reason is that the relative priority (importance) of goals is likely to be different; this necessitates weights for deviation variables. The second reason is that the deviation variables of different goals may not be comparable in their nature (e.g., profit goal in terms of \$/day and total number of chips) and/or magnitude. Mathematically, the objective function of a weighted GP is:

$$f(\mathbf{N}, \mathbf{P}) = \sum_{i=1}^{n} \left[\frac{u_i N_i}{\Phi_i} + \frac{v_i P_i}{\Phi_i} \right] \tag{3.4}$$

Here, n is the number of goals, N_i and P_i are, respectively, negative and positive deviation variables associated with the ith goal, u_i and v_i are, respectively, weights of N_i and P_i and Φ_i is the normalization factor, assumed to be the same for both N_i and P_i of the ith goal.

Weights (u_i and v_i) for those deviation variables that should not be penalized will be zero (i.e., such deviation variables will not appear in the objective function and can take any values). Weights for other deviation variables (that should be penalized) can be based on any suitable method including those covered later in Chapters 5 and 6. They can also be given by decision-makers (e.g., as preference weights in Example 3.1). If priority levels of goals are available, then they can be used to compute these weights, which is illustrated in Example 3.2. Equal weights can also be employed for all deviation variables that should be penalized.

For computing normalization factors, there are several normalization methods (including those in Chapter 4). Some of these are described in the context of GP in Jones and Tamiz (2010). One simple way is to use the given goal values (G_i) for normalizing respective negative/positive deviation variables. This is also illustrated in Example 3.2.

Example 3.2: As a continuation of Example 3.1, formulate the objective function for weighted GP and then solve the GP problem using the Solver tool in Microsoft Excel. State your assumptions, if any.

Solution: As given in Eq. 3.3a, the objective function involves N_1, N_2, P_3 and P_4, which means weights for P_1, P_2, N_3 and N_4 are zero. Let's use preference weights and goal values (as the normalization factors), which are all given in Example 3.1. Given preference weights can be converted into weight fractions as follows:

$$N_1 \text{ weight, } u_1 = \frac{4}{4+4+3+2} = 0.3077 \tag{3.5a}$$

$$N_2 \text{ weight, } u_2 = \frac{4}{4+4+3+2} = 0.3077 \tag{3.5b}$$

$$P_3 \text{ weight}, v_3 = \frac{3}{4+4+3+2} = 0.2308 \tag{3.5c}$$

$$P_4 \text{ weight}, v_4 = \frac{2}{4+4+3+2} = 0.1538 \tag{3.5d}$$

Then, the WS objective function is:

$$f(\mathbf{N}, \mathbf{P}) = \frac{0.3077\ N_1}{600} + \frac{0.3077\ N_2}{60000} + \frac{0.2308\ P_3}{1200} + \frac{0.1538\ P_4}{160} \tag{3.6}$$

Here, normalization factors for N_1, N_2 and P_4 are the respective goal values, whereas the available testing time of 1200 h is used as the normalizing factor for P_3. This objective function and all constraints (Eqs. 3.2a–3.2e) are linear. Decision/deviation variables and their bounds are in Eqs. 3.3b and 3.3c.

The solution of the above LP problem (Eq. 3.6, 3.3b, 3.3c and 3.2a–3.2e) using the Simplex LP option in the Solver tool is given as weighted GP in Table 3.3. According to it, the number of Basic, Advanced and Premium chips to be produced per day is 227, 213 and 160, respectively, which will achieve the first goal of 600 chips/day and the last goal of 160 Premium chips/day. However, profit is only $50,257/day (below the goal of $60,000/day) and additional testing time is 171.8 hours (above the testing time goal of 1200 hours). In short, all goals in a GP problem may not be achieved simultaneously due to the conflicting nature of goals.

TABLE 3.3

Solution of the Example by Weighted, Lexicographic and Min–Max Variants of GP.

Case	Decision Variables	Deviation Variables#	Given Goals with Achieved Values in Brackets
Weighted GP	$x_B = 227$, $x_A = 213$, $x_P = 160$	$N_2 = 9743$, $P_3 = 171.8$	G1 = 600 (600); G2 = 60000 (50257); G3 = 1200 (1371.8); G4 = 160 (160)
Weighted GP (equal weights)	$x_B = 440$, $x_A = 0$, $x_P = 160$	$N_2 = 15920$, $P_3 = 44$	G1 = 600 (600); G2 = 60000 (44080); G3 = 1200 (1244); G4 = 160 (160)
Lexicographic GP	$x_B = 1000$, $x_A = 0$, $x_P = 0$	$N_2 = 6000$, $P_1 = 400$, $P_3 = 500$	G1 = 600 (1000); G2 = 60000 (54000); G3 = 1200 (1700); G4 = 160 (0)
Min–max GP	$x_B = 380$, $x_A = 64$, $x_P = 203$	$N_2 = 8387$, $N_3 = 1.1291$, $N_4 = 1.7452$, $P_1 = 47$, $P_3 = 223.6$, $P_4 = 44.75$	G1 = 600 (647); G2 = 60000 (51613); G3 = 1200 (1422.5); G4 = 160 (203)
Min–max GP (equal weights)	$x_B = 271$, $x_A = 153$, $x_P = 184$	$N_2 = 9299$, $N_3 = 2.98$, $P_1 = 8$, $P_3 = 185.98$, $P_4 = 24.0$	G1 = 600 (608); G2 = 60000 (50701); G3 = 1200 (1383); G4 = 160 (184)

For brevity, only deviation variables having nonzero values are given.

Alternatively, weights for N_1, N_2, P_3 and P_4 can be estimated based on the given reciprocals of priority levels as follows:

$$N_1 \text{ weight}, u_1 = \frac{\frac{1}{1}}{\frac{1}{1}+\frac{1}{1}+\frac{1}{2}+\frac{1}{3}} = 0.3529 \tag{3.7a}$$

$$N_2 \text{ weight}, u_2 = \frac{\frac{1}{1}}{\frac{1}{1}+\frac{1}{1}+\frac{1}{2}+\frac{1}{3}} = 0.3529 \tag{3.7b}$$

$$P_3 \text{ weight}, v_3 = \frac{\frac{1}{2}}{\frac{1}{1}+\frac{1}{1}+\frac{1}{2}+\frac{1}{3}} = 0.1765 \tag{3.7c}$$

$$P_4 \text{ weight}, v_4 = \frac{\frac{1}{3}}{\frac{1}{1}+\frac{1}{1}+\frac{1}{2}+\frac{1}{3}} = 0.1177 \tag{3.7d}$$

For the LP problem with these weights instead of those in Eq. 3.5, the Simplex LP option in the Solver tool gave the same solution obtained earlier (i.e., weighted GP in Table 3.3). The main reason for this is relatively small differences in weight fractions calculated based on preference weights (Eq. 3.5) and those calculated using priority levels (Eq. 3.7).

Let's consider another case of equal weights, that is, N_1 weight = N_2 weight = P_3 weight = P_4 weight = 0.25. The solution of the LP problem with these weights by the Solver tool is presented as weighted GP (Equal Weights) in Table 3.3. This solution is different from that found earlier with two sets of weights (e.g., both profit and additional testing time are less due to a lower weight for N_2 and a higher weight for P_3).

Comments: As illustrated, the optimal solution of the GP problem depends on the weights of deviation variables. Eq. 3.4 indicates that it also depends on the normalization factors, Φ_i.

3.4 Lexicographic Goal Programming

Another distance-based variant of GP is lexicographic GP (also known as preemptive GP), wherein the goals are prioritized into several levels. To solve the GP problem, one or more goals in the top (first) priority level are considered and only the corresponding negative/positive deviation variables are included in the objective function to be minimized. The resulting LP problem is solved. Then, goals in the next priority level are considered and their negative/positive deviations are minimized; in this optimization problem, optimal values of the negative/positive deviation variables minimized in the previous priority level are included as hard constraints. The resulting LP problem is solved. This procedure repeats until all priority levels are covered, or the solution of the LP problem in a priority level is the same as that in the immediately preceding priority level, which indicates that there is only one feasible solution for the LP problem with no scope for further optimization.

Thus, the lexicographic GP method for a problem having, say, p priority levels will require the solution of p or fewer LP problems. Note that p can be equal to the number of goals (n) in applications where there is only one goal in each priority level. In case there are two or more goals in a priority level, the WS of deviation variables must be considered. In the optimal solution by the lexicographic method, only the goals in the higher priority levels are likely to be achieved (and not those in the lower priority levels). The following example illustrates the concepts and procedure of lexicographic GP.

Example 3.3: As a continuation of Example 3.1, formulate the objective function for lexicographic GP and then solve the GP problem using the Solver tool in Microsoft Excel. State your assumptions, if any.

Solution: As given in Eq. 3.3a, the objective function involves N_1, N_2, P_3 and P_4 (and not P_1, P_2, N_3 and N_4). Lexicographic GP requires priority levels. As given in Table 3.2 as part of Example 3.1, there are three priority levels for four goals. Priority level 1 has Goals 1 and 2, whereas each of priority levels 2 and 3 has only one goal, namely, Goals 3 and 4, respectively.

The objective function for the top priority level 1 with Goals 1 and 2 is a function of N_1 and N_2. Since there are two deviation variables to be minimized, their WS can be considered. As given in Table 3.2, preference weights for Goals 1 and 2 are the same (namely, 4). So, equal weight, 0.5, for both N_1 and N_2 is assumed. Recalling that Goal 1 is to manufacture at least 600 chips/day and Goal 2 is to achieve a profit of at least $60,000/day. These two goals have different significance (i.e., cannot be compared) and magnitude. Hence, a normalization factor is required. The normalization factor of a deviation variable is assumed to be equal to the corresponding goal value.

Like in Example 3.2, the WS objective function for priority level 1 with two goals is:

$$f(\mathbf{N}, \mathbf{P}) = \frac{0.5\,N_1}{600} + \frac{0.5\,N_2}{60000} \tag{3.8}$$

Compare this equation with Eq. 3.6 for weighted GP and note the similarities and differences. Decision/deviation variables and their bounds are in Eqs. 3.3b and 3.3c and constraints are in Eqs. 3.2a–3.2e. These constraints and the objective function (Eq. 3.8) are linear. The solution of this LP problem using the Simplex LP option in the Solver tool is $x_B = 1000$, $x_A = x_P = 0$, $N_1 = 0$, $N_2 = 6000$, $N_3 = 0$, $N_4 = 160$, $P_1 = 400$, $P_2 = 0$, $P_3 = 500$ and $P_4 = 0$. Total chips produced per day is 1000 (more than 600 as per Goal 1), profit/day is \$54,000 (less than 60,000 as per Goal 2), additional testing time is 500 hours and lithography time is 800 hours (at the given limit).

Now, proceed to priority level 2 to minimize additional testing time. The objective function for this is simply minimizing P_3 and the additional constraints based on the optimal solution in priority level 1 are $N_1 = 0$ and $N_2 = 6000$ (i.e., deviation variables in the objective function of the previous priority level), besides the constraints in Eqs. 3.2a–3.2e. The additional constraints, $N_1 = 0$ and $N_2 = 6000$, are essential to retain the optimal values of deviation variables penalized in the previous priority level. As before, decision/deviation variables and their bounds are in Eqs. 3.3b and 3.3c. The solution of this LP problem in priority level 2 is the same as that of the LP problem in priority level 1. In other words, additional testing time could not be decreased.

The next/last priority level is to manufacture no more than 160 Premium chips per day, which is already achieved with $x_P = 0$. Setting this aside, the objective function in the last priority level is simply minimizing P_4 and additional constraints based on the optimal values of deviation variables in the objective function in the previous priority levels 1 and 2 are $N_1 = 0$, $N_2 = 6000$ and $P_3 = 500$, besides the constraints in Eqs. 3.2a–3.2e. As before, decision/deviation variables and their bounds are in Eqs. 3.3b and 3.3c. The solution of this LP problem in the last priority level, level 3, is the same as that of the LP problem in priority levels 1 and 2. In other words, optimization in the last priority level, level 3, is not required.

Comments: The optimal solution found by the lexicographic GP is included in Table 3.3. It meets Goal 1 of manufacturing at least 600 chips/day; Goal 2 is not achieved (profit of only \$54000 instead of \$60000); additional testing time (to be minimized as per Goal 3) is substantial at 500 hours; and Goal 4 of manufacturing no more than 160 Premium chips is met. However, producing zero Premium chips may not be acceptable; if so, a suitable constraint on the minimum number of Premium chips should be added in the problem statement. This highlights the importance of correct and comprehensive formulation of any optimization problem (including GP problems).

Changes in priority levels can affect the optimal solution found by the lexicographic GP. This is given as Exercise E3.3 at the end of this chapter. In the above example, Goal 2 of achieving a daily profit of at least \$60,000 is not met. This can be confirmed later by solving an LP problem considering only the daily profit as the single objective (Exercise E3.4 at the end of this chapter).

3.5 Min–Max (Chebyshev) Goal Programming

The third and last variant described in this chapter is the min–max (also known as Chebyshev) GP. Recall that, in the weighted GP, the objective function (Eq. 3.4) is the sum of weighted/normalized deviations of all goals. Instead of this sum, min–max GP uses the maximum of weighted/normalized deviations of all goals as the objective function (Eq. 3.9) to be minimized (i.e., with the intention to minimize the maximum of deviations in all goals).

$$\underset{i=1,2,\ldots,n}{\text{Max}} \left[\frac{u_i N_i}{\Phi_i} + \frac{v_i P_i}{\Phi_i} \right] \tag{3.9}$$

Even in this equation for min–max GP, weighting and normalization factors are employed for the reasons stated earlier. Decision and deviation variables as well as constraints (i.e., Eqs. 3.3b–3.3d) remain the same.

Eq. 3.9 is not linear and its direct use makes the optimization problem nonlinear. To avoid this, a new variable, λ, as the objective to be minimized and the following constraints, which are all linear, are introduced.

$$\left[\frac{u_i N_i}{\Phi_i} + \frac{v_i P_i}{\Phi_i} \right] \leq \lambda \text{ for } i = 1, 2, \ldots, n \tag{3.10}$$

In summary, the LP problem in min–max GP consists of λ as the objective function, those in Eq. 3.1b as well as λ as decision/deviation variables, bounds as in Eq. 3.1c and Eqs. 3.2a–3.2e and Eq. 3.10 as constraints. The following example illustrates the min–max GP problem for an application and its solution.

Example 3.4: Formulate the complete LP problem for the application of min–max GP to Example 3.1. Then, solve it using the Solver tool in Microsoft Excel. State your assumptions, if any.

Solution: Recall and review Example 3.1 and the formulated GP problem in Eq. 3.3. The deviation variables to be penalized are N_1, N_2, P_3 and P_4 and their normalization factors are 600, 60000, 1200 and 160, respectively (Eq. 3.6). Their weights, based on given preference weights, are those computed in Eq. 3.5. The complete min–max GP problem is as follows:

$$\text{Minimize} \quad \lambda \tag{3.11a}$$

$$\text{With respect to} \quad \lambda, N_1, N_2, N_3, N_4, P_1, P_2, P_3, P_4, x_B, x_A, x_P \tag{3.11b}$$

Subject to

$$\lambda, N_1, N_2, N_3, N_4, P_1, P_2, P_3, P_4 \geq 0, \text{ and } x_B, x_A, x_P \geq 0 \text{ and Integer} \quad (3.11c)$$

$$x_B + x_A + x_P + N_1 - P_1 = 600 \quad (3.11d)$$

$$54x_B + 83x_A + 127x_P + N_2 - P_2 = 60,000 \quad (3.11e)$$

$$1.7x_B + 2.3x_A + 3.1x_P + N_3 - P_3 = 1200 \quad (3.11f)$$

$$x_P + N_4 - P_4 = 160 \quad (3.11g)$$

$$0.8x_B + 1.4x_A + 2.0x_P \leq 800 \quad (3.11h)$$

$$\frac{0.3077\ N_1}{600} \leq \lambda \quad (3.11i)$$

$$\frac{0.3077\ N_2}{60000} \leq \lambda \quad (3.11j)$$

$$\frac{0.2308\ P_3}{1200} \leq \lambda \quad (3.11k)$$

$$\frac{0.1538 P_4}{160} \leq \lambda \quad (3.11l)$$

Note that Eqs. 3.11d–3.11h are, respectively, Eqs. 3.2a–3.2e developed in Example 3.1. Eqs. 3.11i–3.11l are the additional constraints introduced for minimizing the maximum of deviations of four goals (through linear equations); these are only for the deviation variables to be penalized.

The LP problem in Eq. 3.11 can be solved using the Simplex LP option in the Solver tool of Microsoft Excel. The optimal solution found is given in the row of min–max GP in Table 3.3. In this, only Goal 1 (i.e., more than 600 chips/day) is achieved and Goals 2–4 are not met: Profit/day is $51,613 (less than Goal 2 of at least $60,000), additional testing required is 222.5 hours and 43 more Premium chips are produced.

Comments: Like weighted GP, weights and normalization factors affect the solution by the min–max GP as well. For example, consider equal weights of 0.25 instead of 0.3077, 0.3077, 0.2308 and 0.1538 in Eqs. 3.11i–3.11l. The solution obtained by solving the LP problem in this case is given as min–max GP (Equal Weights) in Table 3.3. In this, only Goal 1 is achieved, whereas the other three goals (profit of at least $60,000/day, avoid additional testing and produce no more than 160 Premium chips) are not met; this is like the solution obtained earlier with weights based on preference weights, but the deviations from the goals are different.

3.6 Merits of, Limitations of and Guidelines for Goal Programming

As seen in the above sections, GP solves an optimization problem having multiple objectives (criteria) to find an optimal solution. Thus, it belongs to MCDM, specifically to the MODM group of MCDM (Section 1.1) and it is also a method for solving MOO problems. Hence, it shares some merits (e.g., being able to handle many criteria) and limitations (e.g., subjectivity) of MCDM. Merits and limitations of GP are outlined below. Then, guidelines to avoid pitfalls in GP are summarized.

Merits of GP: First, the concepts of goals and their achievement in GP are intuitive and easy to understand. Second, it gives flexibility to set goals that are desired but not compulsory to achieve. Third, via goals, GP tries to find satisfactory (and not optimal) values for criteria. Fourth, decision-makers can give their priority levels, preferences and/or weights in formulating GP problems. Fifth, there are several variants (e.g., linear, fuzzy, integer/binary and fractional) of GP for decision-makers to choose from as per their emphasis and application. Among them, linear GP is based on established LP, known to engineers and scientists and whose solution is guaranteed (unless there is no feasible solution). For example, lexicographic GP is suitable if there are priority levels for goals, whereas min–max GP is preferable to reduce the maximum deviation from the given goal values. Sixth, GP can be used with any number of goals, decision variables and system constraints and its solution in the case of linear equations is computationally undemanding. Seventh, GP can be easily extended to applications involving nonlinear constraints and criteria; although this is not covered in some books on GP (e.g., Schniederjans, 1994; Jones and Tamiz, 2010), Padmanabhan and Vrat (1990) reported an application of nonlinear GP to multi-item inventory systems under resource constraints. Finally, GP has numerous applications in diverse fields.

Limitations of GP: First, giving acceptable values, priority levels, preferences and/or weights for all goals can be challenging to decision-makers, although it is not so difficult for goals like available resources, costs and profit when there is in-depth knowledge of the application. Second, GP finds only one solution (and not many or all) Pareto-optimal solutions; sometimes, the solution found may not be Pareto-optimal. Third, in lexicographic GP, there may be only one solution for the LP problem after the first few priority levels and so subsequent priority levels have no effect (referred to as degeneracy). Fourth, a nonlinear GP with nonlinear constraints is challenging to solve and computationally demanding. In such cases, one may have to employ the methods covered in Chapter 2.

Jones and Tamiz (2010) listed rules or guidelines to avoid pitfalls and overcome limitations in formulating and solving GP problems. First, always include both negative and positive deviation variables in each goal equation (Eq. 3.1d). Second, penalize only the unwanted (and not all) deviations. Third, consider priority levels, preferences and/or weights as initial estimates (and not definite values). This means always perform a sensitivity analysis of the optimal solution for these inputs. Fourth, choose an appropriate GP variant based on the problem nature and decision-makers' aims. Fifth, linear GP variants described in this chapter and some others will require normalization of deviation variables; use an appropriate normalization method and later study its effect on the solution. Sixth, in lexicographic GP, do not use an excessive number of priority levels and optimum (ideal target) values for goals. The suggested number of priority levels is 3 or 4. Finally, check whether the solution found by GP is Pareto-optimal or not. If required, correct it to be Pareto-optimal; for details on these, see Chapter 6 in Jones and Tamiz (2010).

3.7 Goal Programming for Alternatives–Criteria Matrix Data

Recalling from Chapter 1, the ACM represents a table with alternatives listed in rows and criteria in columns. One example is the forklift selection in Table 1.1 and many examples with ACM data are given in Appendix A. In these examples, alternatives and criteria values for them are given; there are no decision variables and no need to solve any optimization problem (such as Eq. 3.1). Many MCDM methods for solving such examples are covered in Chapters 7–9.

Here, we consider the application of GP variants described in this chapter to ACM data to choose one of the alternatives. The approach and concepts of GP can be adopted for ACM data without introducing negative/positive deviation variables. The generic steps are outlined in this section using the following notation.

- $i \in \{1,2,\ldots,m\}$ and $j \in \{1,2,\ldots,n\}$, where m is the number of rows (i.e., alternatives) and n is the number of columns (i.e., criteria) in the ACM.
- f_{ij} is the value of the jth criterion for the ith alternative in the original ACM.

- F_{ij} is the value of the jth criterion for the ith alternative in the normalized ACM.
- D_{ij} is the value of deviation from the goal of jth criterion (G_j) for the ith alternative in the normalized ACM.
- G_j is the goal of criterion $j \in \{1,2,\ldots,n\}$.
- v_{ij} is the value of the jth criterion for the ith alternative in the weighted normalized ACM.
- w_j is the weight assigned to the jth criterion and $\sum_{j=1}^{n} w_j = 1$.
- ws_i is WS for alternative i.

Additionally, ACM data in Table 3.4 are employed to illustrate the steps in the application of GP methods to it. This ACM is made up of four alternatives (i.e., $m = 4$ and $i \in \{1,2,3,4\}$) and three criteria (i.e., $n = 3$ and $j \in \{1,2,3\}$). Out of the three criteria, C1 and C2 are benefit criteria to be maximized, while C3 is a cost criterion to be minimized. For illustration, weights for the three criteria are assumed as $w_1 = 0.25$, $w_2 = 0.33$ and $w_3 = 0.42$. All four alternatives (A1, A2, A3 and A4) are non-dominated, which means that, among the alternatives, the improvement in one criterion will inevitably result in degradation in at least one other criterion (i.e., a trade-off among the criteria).

Step 1. Normalize the original ACM with m rows (i.e., alternatives) and n columns (i.e., criteria) using the max normalization method (which and other methods are described in Chapter 4). For a maximization criterion, the normalized value, F_{ij}, according to this method, is given by:

$$F_{ij} = \frac{f_{ij}}{\max\limits_{k \in \{1,2,\ldots,m\}} f_{kj}} \tag{3.12}$$

TABLE 3.4

ACM Employed for Illustration of GP (C1 and C2 for Maximization; C3 for Minimization).

Alternatives	C1 (Max)	C2 (Max)	C3 (Min)
A1	0.93	600	8.25
A2	0.51	700	6.33
A3	0.77	500	3.16
A4	0.82	400	2.98
Weight	0.25	0.33	0.42

On the other hand, F_{ij} for a minimization criterion is:

$$F_{ij} = \frac{\min_{k\in\{1,2,\ldots,m\}} f_{kj}}{f_{ij}} \tag{3.13}$$

The above normalization ensures that all minimization criteria are transformed to the maximization type (i.e., a higher F_{ij} is better and preferred). Earlier in Eq. 3.4, goal value was used as the normalization factor, mainly due to the unavailability of the maximum/minimum of a criterion, whereas these are readily available from the ACM data itself.

NUMERICAL CALCULATIONS (USING ACM FROM TABLE 3.4)

For instance, for $i=1$ and $j=2$: $F_{12} = \frac{f_{12}}{\max_{k\in\{1,2,3,4\}} f_{k2}} = \frac{600}{700} = 0.8571$.

Similar calculations are performed for all other values in the ACM across the four alternatives and three criteria. The complete results, forming the normalized ACM, are shown in Table 3.5.

TABLE 3.5

Normalized ACM for Illustration of GP.

Alternatives	C1	C2	C3
A1	1	0.8571	0.3612
A2	0.5484	1	0.4708
A3	0.8280	0.7143	0.9430
A4	0.8817	0.5714	1

Step 2. Choose a suitable goal, G_j, for each criterion, $j \in \{1,2,\ldots,n\}$. For the normalized ACM, this can be set at a fraction (e.g., 0.8, 0.9, or 0.95) of the maximum (i.e., 1.0); this fraction need not be the same for all criteria. Then, compute the deviation of each criterion from its goal as follows:

$$D_{ij} = G_j - F_{ij}, \quad i \in \{1,2,\ldots,m\}, j \in \{1,2,\ldots,n\} \tag{3.14}$$

The above equation is for all criteria and all alternatives. Deviation for a criterion of an alternative can be positive (indicating underachievement of the goal) or negative (meaning overachievement of the goal). Obviously, overachievement (i.e., negative value) is preferred.

NUMERICAL CALCULATIONS

Assume the goal, G_j, is 0.8 for all criteria. Then, for $i=1$ and $j=2$: $D_{12}=G_2-F_{12}=0.8-0.8571=-0.0571$. Similar calculations are performed for all other values in the normalized ACM across the four alternatives and three criteria. The complete results are in Table 3.6.

TABLE 3.6

Deviations from Goals (in Columns 2, 3 and 4) and Weighted Sum (in the Last Column), for all Alternatives.

	Deviation in			
Alternatives	**C1**	**C2**	**C3**	**Weighted Sum**
A1	−0.2000	−0.0571	0.4388	0.115434
A2	0.2516	−0.2000	0.3292	0.135178
A3	−0.0280	0.0857	−0.1430	−0.03878
A4	−0.0817	0.2286	−0.2000	−0.029

Step 3. Weighted GP can be applied to the deviations in Table 3.6 along with the criteria weights in Table 3.4. The WS for alternative i (ws_i) is computed as follows:

$$ws_i=\sum_{j=1}^{n}\left[w_j D_{ij}\right] \tag{3.15}$$

The above equation is applied for every alternative (i.e., $i=1,2,3,4$). WS of an alternative can be negative due to negative values for deviations (as shown in Table 3.6). The alternative with the lowest WS is the top-ranked and recommended. Note that a negative deviation, indicating overachievement of a goal, reduces the WS; this is justified since the top-ranked alternative is the one with the lowest WS.

NUMERICAL CALCULATIONS

The given weights are $w_1=0.25$, $w_2=0.33$ and $w_3=0.42$. For instance, for $i=3$:

$$ws_3=0.25\times(-0.0280)+0.33\times(0.0857)+0.42\times(-0.1430)=-0.03878$$

Similar calculations are performed for all other alternatives. WS results are given in the last column of Table 3.6. Accordingly, alternative 3 with the lowest WS is the top-ranked and recommended to decision-makers.

Lexicographic GP can also be applied to deviations from goals in Table 3.6. This requires priority levels for the goals and simple logic, as illustrated below.

NUMERICAL CALCULATIONS

For applying lexicographic GP, assume top priority level 1 is maximizing C1 and the next priority level 2 is minimizing C3. Recall that negative deviation means overachievement of the goal. As can be seen in Table 3.6, in priority level 1, alternatives 1, 3 and 4 exceed the goal for C1 and so all of them are acceptable. Proceeding to priority level 2, alternative 4 (among alternatives 1, 3 and 4) exceeds the goal for C3 by the most; in fact, it has the lowest value for C3 (Table 3.4). Hence, alternative 4 is recommended as per the lexicographic GP.

Analogously, min–max GP can be applied to the deviations from goals in Table 3.6, identifying the highest deviation (among deviations in all goals) for each alternative and then finding the alternative that has the lowest value among these highest deviations. Note that the deviations should be considered as they are (and not absolute values); the highest deviation for each alternative is very likely to be positive.

NUMERICAL CALCULATIONS

Inspection of all deviations in Table 3.6 shows that the maximum deviation for alternatives 1, 2, 3 and 4 is, respectively, 0.4388, 0.3292, 0.0857 and 0.2286. Of these, the minimum deviation is 0.0857 for alternative 3. Hence, alternative 3 is recommended as per min–max GP.

Comments: The alternative recommended by weighted, lexicographic and min–max GP can be the same or different. In the illustrated example, both weighted and min–max GP gave alternative 3, whereas lexicographic GP recommended alternative 4. This can change with different goals, weights, normalization methods and priority levels.

3.8 Comparison of Goal Programming with Related Methods

GP, ranking/MCDM methods (introduced in Chapter 1) and MOO (covered in Chapter 2) all fall under the category of optimization, going by its dictionary meaning of finding the best or most effective solution. Hence, they have similarities as well as differences. Earlier, in Section 2.6, MCDM and MOO are compared. Here, in this section, GP is first compared with ranking/MCDM methods and then with MOO.

The general form of a GP problem (Eq. 3.1) is an optimization problem with deviation and decision variables (having bounds on them), an objective function of deviation variables, soft constraints for goals (related to criteria) and system/hard constraints. GP methods are for solving this problem to find an optimal solution that overachieves or underachieves the goals. Ranking/MCDM methods use given ACM data to rank the alternatives and/or select one alternative (Nabavi et al., 2023; Wang et al., 2024). As described in the previous section, GP methods can be applied to ACM data to rank and select one alternative. Thus, both GP and ranking methods optimize to find the best or most effective solution/alternative. Further, they require a normalization factor and a weight for each criterion (goal).

The differences between GP and ranking/MCDM methods include the following. Compared to GP problems, ACM data used in ranking methods do not involve deviation/decision variables, soft/hard constraints, goals and a new objective function of deviation variables. In fact, the ranking of ACM data does not involve formulating or solving a structured mathematical optimization problem (such as Eq. 3.1 with deviation variables, constraints and an objective function); consequently, it does not require complex iterative algorithms (Wang et al., 2021). The solution of linear GP problems is by an LP method such as the Simplex method and that of nonlinear GP problems requires an iterative or numerical method such as SQP or GAs; manual calculations for these methods are tedious and so they require a computer program. The solution of a given/formulated GP problem is the same irrespective of the numerical method employed to solve the problem. Obviously, the solution changes with the formulated problem. On the other hand, there are more than 200 ranking/MCDM methods (Baydaş et al., 2024) and the ranking of given ACM data can be different depending on the ranking method employed (Wang et al., 2023).

Let's proceed to compare GP with MOO covered in Chapter 2. GP problems (Eq. 3.1) and MOO problems (Eq. 2.1) are mathematically similar. Common features in these problems are decision variables (with bounds), constraints and objective function(s). Both these can be solved by the same computational methods (e.g., linear problems by the Simplex method and nonlinear problems by SQP or GAs), which are all iterative; their solution does not vary with the computational method (except for minor variations due to the use of random numbers in methods such as GAs).

The differences between GP and MOO problems include the following. GP problems are often presented as linear type, whereas MOO problems are presented as nonlinear type, although they are not restricted to either type. The goals, deviation variables and soft constraints occurring in GP problems are absent in MOO problems. Only one objective function is involved in a GP problem and it is a function of deviation variables (and not decision variables). On the other hand, there are multiple objectives (criteria) in an MOO problem and they are functions of some or all decision variables (Wang et al., 2025). Generally, there is only one optimal solution to the GP problem, whereas an MOO problem has many solutions, known as Pareto-optimal solutions.

3.9 Python Programs for Solving Goal Programming Problems

There are off-the-shelf Python libraries for solving GP problems. For instance, PuLP (Python Linear Programming), an open-source Python library developed by Mitchell et al. (2011), is for modeling linear optimization problems. It allows coding LP and mixed-integer linear programming (MILP) models entirely within the Python environment; users can define decision variables, deviation variables, objective functions and constraints in a flexible manner. PuLP library serves primarily as a coding tool; it internally generates LP or MILP formulations that are solved by external solvers such as the Computational Infrastructure for Operations Research (COIN-OR)'s Branch and Cut (CBC) solver written in C++, GNU Linear Programming Kit (GLPK) and others. Due to its ease of use, seamless integration with Python and compatibility with a range of solvers, PuLP is widely used for prototyping and solving practical optimization problems (Mitchell et al., 2011; Jha et al., 2023).

GP problems can be effectively coded and solved using PuLP by representing the problem as an LP model with deviation variables. As discussed in the previous sections, in GP, decision-makers specify goals for multiple, often conflicting, criteria. To model this, positive and negative deviation variables are introduced for each goal to quantify overachievement or underachievement. These deviations are then minimized according to a chosen technique (e.g., WS, lexicographic and min–max). The coding flexibility of PuLP enables users to construct a GP problem accordingly. By defining relevant decision variables, constraints and an objective function that minimizes the deviations, the GP problem can be solved using PuLP.

For example, using the PuLP library, Table 3.7 presents the complete Python code and output for solving the weighted GP problem formulated in Section 3.3 for the semiconductor manufacturing plant that produces three types of memory chips (Example 3.1). The results obtained using Python PuLP are the same as those previously reported in Section 3.3 using Excel's Solver tool (Table 3.3). Specifically, the optimal daily production quantities

TABLE 3.7

Python Code and Output for Solving Weighted GP Problem Using PuLP Library (Version 3.2.1).

```
# pip install pulp # uncomment this line to install the pulp library
import pulp

# 1) Define the GP problem
prob = pulp.LpProblem("Semicon_GP_Weighted", pulp.LpMinimize)

# 2) Decision vars (integer, ≥0)
xB = pulp.LpVariable('xB', lowBound=0, cat='Integer')
xA = pulp.LpVariable('xA', lowBound=0, cat='Integer')
xP = pulp.LpVariable('xP', lowBound=0, cat='Integer')

# 3) Deviation vars (continuous, ≥0)
N1 = pulp.LpVariable('N1', lowBound=0)
P1 = pulp.LpVariable('P1', lowBound=0)
N2 = pulp.LpVariable('N2', lowBound=0)
P2 = pulp.LpVariable('P2', lowBound=0)
N3 = pulp.LpVariable('N3', lowBound=0)
P3 = pulp.LpVariable('P3', lowBound=0)
N4 = pulp.LpVariable('N4', lowBound=0)
P4 = pulp.LpVariable('P4', lowBound=0)

# 4) Goal constraints
prob += ( xB +    xA +    xP + N1 - P1 == 600 ), "Goal_1_chip_count"
prob += ( 54*xB + 83*xA +127*xP + N2 - P2 == 60000 ), "Goal_2_profit"
prob += ( 1.7*xB +2.3*xA + 3.1*xP + N3 - P3 == 1200 ), "Goal_3_test_time"
prob += ( xP + N4 - P4 == 160 ), "Goal_4_premium_count"

# 5) Hard/system constraint
prob += ( 0.8*xB + 1.4*xA + 2.0*xP <= 800 ), "Photo_time_limit"

# 6) Weights for the deviation variables
w1 = 0.3077
w2 = 0.3077
w3 = 0.2308
w4 = 0.1538

# 7) Obj: minimize weighted normalized sum of "undesirable" deviations
prob += (
    w1 * N1/600 +
    w2 * N2/60000 +
    w3 * P3/1200 +
    w4 * P4/160
), "Total_Weighted_Deviation"

# 8) Solve
prob.solve(pulp.PULP_CBC_CMD(msg=False))
```

(*Continued*)

TABLE 3.7 *(Continued)*

Python Code and Output for Solving Weighted GP Problem Using PuLP Library (Version 3.2.1).

```
# 9) Report decision, deviation vars and goals
print("Final Solution:")
for v in [xB, xA, xP, N1, N2, N3, N4, P1, P2, P3, P4]:
    print(f"{v.name:>4} = {v.value():>6.1f}")
print(f"  Obj = {pulp.value(prob.objective):.5f}")

# 10) Print each goal's value
chips     = xB + xA + xP
profit    = 54*xB + 83*xA + 127*xP
test_time = 1.7*xB + 2.3*xA + 3.1*xP
premium   = xP
print(f"  Goal 1 Total Chips/day   = {pulp.value(chips):.1f}")
print(f"  Goal 2 Profit/day        = {pulp.value(profit):.1f}")
print(f"  Goal 3 Testing Time      = {pulp.value(test_time):.1f}")
print(f"  Goal 4 Premium Chips/day = {pulp.value(premium):.1f}")

# -----------------
# Final Solution:
#   xB =  227.0
#   xA =  213.0
#   xP =  160.0
#   N1 =    0.0
#   N2 = 9743.0
#   N3 =    0.0
#   N4 =    0.0
#   P1 =    0.0
#   P2 =    0.0
#   P3 =  171.8
#   P4 =    0.0
#   Obj = 0.08301
#   Goal 1 Total Chips/day   = 600.0
#   Goal 2 Profit/day        = 50257.0
#   Goal 3 Testing Time      = 1371.8
#   Goal 4 Premium Chips/day = 160.0
# -----------------
```

are 227 Basic chips, 213 Advanced chips and 160 Premium chips. The corresponding outcomes for goals are: Goal 1 – total chip production of 600 chips/day; Goal 2 – profit of $50,257/day; Goal 3 – testing time of 1371.8 hours; and Goal 4 – production of 160 Premium chips/day.

In addition to PuLP, several other open-source Python libraries can be used to solve GP problems; they include CVXPY (Diamond and Boyd, 2016), Python-MIP (Santos and Toffolo, 2020) and Pyomo (Bynum et al., 2021). Further, Chapter 6 in the book by Papathanasiou and Ploskas (2018) describes weighted, lexicographic and min–max GPs and illustrates their implementation in Python via a numerical example.

3.10 Summary

Main points in this chapter on GP problems and their solutions are as follows:

1. A GP problem is an optimization problem involving goals, deviation variables (from goals), decision variables, bounds, soft and hard constraints and an objective function of deviation variables only. It is generally presented as a linear problem, although it can be nonlinear.
2. A GP problem requires goals, one for each criterion and it allows underachievement or overachievement of each goal.
3. There are many variants of GP such as weighted GP, lexicographic GP and min–max GP, described and illustrated in this chapter. These require normalization factors and weights for goals (criteria).
4. Decision-makers must provide values of goals, normalization factors and weights for formulating a GP problem. This may or may not be easy, depending on the application and background of decision-makers.
5. Solution of a linear GP problem requires the solution of one or more LP problems. It depends on the given values for goals, normalization factors and weights as well as the GP variant employed. Hence, both problem formulation and sensitivity analysis are essential.
6. A GP problem can be solved using the built-in Solver tool within Microsoft Excel or using Python open-source libraries.
7. Although popular, GP has several merits and limitations; guidelines for its successful use are outlined.
8. As illustrated in Section 3.7, GP variants can be applied to ACM data for ranking and selecting an alternative.
9. GP, ranking/MCDM and MOO methods find the optimum or most effective solution and so they are all for optimization. Hence, there are similarities and differences among these types of methods.
10. An open-source Python library: PuLP, illustrated for a weighted GP in Section 3.9, is useful for solving linear GP problems.

References

Baydaş, M., Kavacık, M., & Wang, Z. (2024). Interpreting the determinants of sensitivity in mcdm methods with a new perspective: An application on e-scooter selection with the probid method. *Spectrum of Engineering and Management Sciences*, *2*(1), 17–35.

Bynum, M. L., Hackebeil, G. A., Hart, W. E., Laird, C. D., Nicholson, B. L., Siirola, J. D., Watson, J.P., & Woodruff, D. L. (2021). *Pyomo-optimization modeling in python*, Springer Optimization and Its Applications (Vol. 67). Berlin, Heidelberg: Springer.

Charnes, A., & Cooper, W. W. (1961). *Management models and industrial applications of linear programming*. New York: Wiley.

Charnes, A., Cooper, W. W., & Ferguson, R. (1955). Optimal estimation of executive compensation by linear programming. *Management Science, 1*, 138–151.

Colapinto, C., Jayaraman, R., & Marsiglio, S. (2015). Multi-criteria decision analysis with goal programming in engineering, management and social sciences: A state-of-the art review. *Annals of Operations Research*, Online First, 1–34.

Diamond, S., & Boyd, S. (2016). CVXPY: A python-embedded modeling language for convex optimization. *Journal of Machine Learning Research, 17*(83), 1–5.

Edgar, T. F., Himmelblau, D. M., & Lasdon, L. S. (2001). *Optimization of chemical processes* (2nd ed.). McGraw-Hill.

Jha, A. K., Sharma, G. S., & Kaushik, N. (2023). Modeling of transport problem in linear programming with python (PULP). *SAMRIDDHI: A Journal of Physical Sciences, Engineering and Technology, 15*(1), 104–108.

Jones, D., & Tamiz, M. (2010). *Practical goal programming*. Springer.

Mitchell, S., O'Sullivan, M., & Dunning, I. (2011). *Pulp: A linear programming toolkit for python*. The University of Auckland, Auckland, New Zealand, 65, 25.

Nabavi, S. R., Jafari, M. J., & Wang, Z. (2023). Deep learning aided multi-objective optimization and multi-criteria decision making in thermal cracking process for olefines production. *Journal of the Taiwan Institute of Chemical Engineers, 152*, 105179.

Padmanabhan, G., & Vrat, P. (1990). Analysis of multi-item inventory systems under resource constraints: A non-linear goal programming approach. *Engineering Costs and Production Economics, 20*, 121–127.

Papathanasiou, J., & Ploskas, N. (2018). *Multiple criteria decision aid: Methods, examples and python implementation*. Springer.

Santos, H. G., & Toffolo, T. (2020). Mixed integer linear programming with python. *Computational Infrastructure for Operations Research (COIN-OR)*. https://app.readthedocs.org/projects/python-mip/downloads/pdf/latest/ (accessed in July 2025).

Schniederjans, M. J. (1994). *Goal programming: Methodology and applications*. Kluwer Academic Publishers.

Wang, Y., Wang, Z., & Wu, Z. (2025). Multi-objective optimal control of nonlinear processes using reinforcement learning with adaptive weighting. *Computers & Chemical Engineering*, 109206.

Wang, Z., Baydaş, M., Stević, Ž., Özçil, A., Irfan, S. A., Wu, Z., & Rangaiah, G. P. (2023). Comparison of fuzzy and crisp decision matrices: An evaluation on PROBID and sPROBID multi-criteria decision-making methods. *Demonstratio Mathematica, 56*(1), 20230117.

Wang, Z., Nabavi, S. R., & Rangaiah, G. P. (2024). Multi-criteria decision making in chemical and process engineering: Methods, progress and potential. *Processes, 12*(11), 2532.

Wang, Z., Rangaiah, G. P., & Wang, X. (2021). Preference ranking on the basis of ideal-average distance method for multi-criteria decision-making. *Industrial & Engineering Chemistry Research, 60*(30), 11216–11230.

Exercises

E3.1 State whether each of the following statements is correct or incorrect.

1. GP problem is essentially a mathematical optimization problem with an objective function, deviation/decision variables, bounds and constraints.
2. GP problems are always linear.
3. A GP problem can have both soft (goal) constraints and hard (system) constraints.
4. Solution of GP problems does not require both the normalization factor and weight for each criterion/goal.
5. In lexicographic GP, goals are handled considering one priority level at a time and lower-level (less important) goals are optimized only after higher-level (more important) goals are satisfied as much as possible.
6. In min–max GP, the objective is to minimize the sum of weighted deviations from all goals.
7. The solution of a GP problem by different variants of GP will be the same.
8. The solution of a GP problem depends on the normalization factors and weights.
9. Both GP and MOO problems involve deviation variables besides decision variables.
10. A given GP problem has only one optimal solution, whereas a given MOO problem has many optimal solutions.
11. The concept of goals can be employed for ranking the given alternatives and their values for the criteria.

E3.2 A logistics company distributes a product from two distribution centers to three retail stores. The available supplies at Distribution Centers AA and BB are 1000 units and 1500 units, respectively. The demands at four retail stores are 600, 900, 750 and 800 units, respectively. The net revenue per unit shipped from distribution center i to retail store j is shown in the table below:

Distribution Center	Store 1	Store 2	Store 3	Store 4
AA	6.5	9.0	8.5	7.5
BB	6.0	8.0	7.0	7.0

The company has established the following shipping goals (in decreasing order of priority/importance):

1. Completely fulfill the demand at Store 2.
2. Maximize the total revenue.
3. Satisfy at least 80% of the demand at Store 3.
4. Ensure that at least 650 units are shipped from Distribution Center AA.

Formulate a GP problem for the above application. State your assumptions, if any. Then, solve the GP problem using weighted, lexicographic and/or min–max variants and an LP solver (in either Microsoft Excel or Python).

E3.3 Consider Example 3.1 with the changed priority levels and preference weights for four goals, given below. Solve this new case by weighted GP, lexicographic GP and min–max GP and compare the solution found by each GP variant with the corresponding solution in Examples 3.2–3.4. Do priority levels and preference weights affect the optimal solution by a GP variant?

Goal	Priority Level	Preference Weight
Goal 1 – Achieve a profit of at least \$60,000/day	1	5
Goal 2 – Manufacture at least 600 chips/day	2	4
Goal 3 – Minimize additional testing time required	3	3
Goal 4 – Manufacture no more than 160 Premium chips per day	3	2

E3.4 Consider Example 3.1 but with <u>no</u> goals, priority levels and reference weights in Table 3.2. Assume additional testing time is zero. Formulate <u>an LP problem</u> for maximizing the profit. In this problem, what are the decision variables, constraints and objective function? Will there be any deviation variables and soft constraints? Solve the formulated LP problem using the Solver tool in Microsoft Excel or a suitable Python library. In the optimal solution found, what are the maximum profit and optimal values of the decision variables? How many Premium chips and how many total chips are manufactured per day? Briefly discuss the ways to further increase the optimal profit; make suitable assumptions for this discussion.

E3.5 As a continuation of Example 3.1, formulate the objective function for weighted GP and then solve it. For this, assume weights are those given in Eq. 3.5 and the normalization factor for N_1, N_2, P_3 and P_4 is the same, equal to 600 + 60,000 + 1,200 + 160 = 61,960, which is based on the sum normalization described in Chapter 4 and values of goals including available testing time. Compare the solution obtained using this normalization factor with that found in Example 3.2.

E3.6 Apply weighted, lexicographic and min–max variants of GP to the following ACM data. Criteria weights are in the last row.

Alternatives	C1 (Max)	C2 (Min)	C3 (Max)
A1	0.185	2.33	454
A2	0.317	1.08	298
A3	0.555	6.45	174
A4	0.731	8.88	849
A5	0.948	7.39	517
Weight	0.25	0.4	0.35

E3.7 Apply weighted, lexicographic and min–max variants of GP to the following ACM data. Criteria weights are in the last row.

Alternatives	C1 (Max)	C2 (Max)	C3 (Max)	C4 (Min)
A1	0.78	117	6.2	35
A2	0.65	150	5.8	40
A3	0.83	110	4.9	30
Weight	0.30	0.25	0.20	0.25

4

Normalization Methods

4.1 Overview

Multi-criteria decision-making (MCDM) typically considers multiple criteria simultaneously to rank the alternatives and/or choose one of the alternatives (Wang et al., 2023). Generally, multiple criteria are distinct in terms of their significance and magnitude (Nabavi et al., 2024). This can be seen, for example, from the forklift selection application in Table 1.1, wherein the criteria are purchase cost, maintenance cost, fuel consumption, *warranty period, *service network, *availability of spare parts, *maximum load capacity, *maximum lifting capacity, *movement speed and *speed of lifting/lowering loads. Of these, the first three are to be minimized (i.e., cost criteria), whereas the remaining seven criteria (with * in the beginning) are to be maximized (i.e., benefit criteria). Hence, normalization of criteria values, say, to the range of 0–1, is required in many ranking/MCDM methods (presented in Chapters 7–9) and in weighted goal programming (GP) (Chapter 3) and weighted sum method (WSM) for solving multi-objective optimization (MOO) problems (Chapter 2).

This chapter describes four popular normalization methods used in MCDM, namely, max, max–min, sum and vector normalization methods. The underlying equations in these methods are relatively simple and straightforward. Even then, a description of each of these methods in this chapter is followed by a numerical example. Accordingly, the remaining sections of this chapter are as follows:

Section 4.2	Max Normalization
Section 4.3	Max–Min Normalization
Section 4.4	Sum Normalization
Section 4.5	Vector Normalization
Section 4.6	Discussion
Section 4.7	Numerical Difficulties
Section 4.8	Summary

DOI: 10.1201/9781003530435-4

The learning outcomes of this chapter are to (1) describe the principles of the normalization methods covered and (2) apply any of these methods to ACM of an application.

Unless otherwise stated, the notation used throughout this chapter is as follows:

- $i \in \{1,2,\ldots,m\}$ and $j \in \{1,2,\ldots,n\}$, where m is the number of rows (i.e., alternatives) and n is the number of columns (i.e., criteria) in the ACM.
- f_{ij} is the value of the jth criterion for the ith alternative in the original ACM.
- F_{ij} is the normalized value of f_{ij}.

Additionally, ACM presented in Table 4.1 is employed to illustrate the normalization methods covered in this chapter. This ACM is made up of five alternatives (i.e., $m = 5$ and $i \in \{1,2,3,4,5\}$) and three criteria (i.e., $n = 3$ and $j \in \{1,2,3\}$). Out of the three criteria (C1, C2 and C3), C1 and C3 are benefit criteria to be maximized, whereas C2 is a cost criterion to be minimized. All five alternatives (A1, A2, A3, A4 and A5) are non-dominated, which means that the improvement in one criterion will inevitably result in degradation in at least one other criterion (i.e., a trade-off among the criteria).

4.2 Max Normalization

In the max normalization, probably the simplest of many methods available for normalization, values of a benefit criterion for all m alternatives are normalized by:

$$F_{ij} = \frac{f_{ij}}{\max_{k \in \{1,2,\ldots,m\}} f_{kj}} \tag{4.1a}$$

TABLE 4.1

ACM Employed for Illustration of Normalization Methods (C1 and C3 for Maximization; C2 for Minimization).

Alternatives	C1 (Max)	C2 (Min)	C3 (Max)
A1	0.185	2.33	454
A2	0.317	1.08	298
A3	0.555	6.45	174
A4	0.731	8.88	849
A5	0.948	7.39	517

Here, the denominator is the maximum of the values of the benefit criterion across all m alternatives. Eq. 4.1a is employed for every benefit criterion in ACM. On the other hand, values of a cost criterion for all m alternatives are normalized by a different equation as follows:

$$F_{ij} = \frac{\min_{k \in \{1,2,\ldots,m\}} f_{kj}}{f_{ij}} \tag{4.1b}$$

Here, the numerator is the minimum of values of the cost criterion across all alternatives. Eq. 4.1b is used for every cost criterion in ACM. After max normalization, the maximum and minimum of any criterion are exactly 1.0 and less than 1.0, respectively. Moreover, values of a cost criterion become like those of a benefit criterion (i.e., larger values are preferred). For example, max normalization is employed in the SAW method of MCDM (Wang et al., 2022), as described in Chapter 7.

NUMERICAL CALCULATIONS (USING ACM FROM TABLE 4.1)

For instance, in the case of the maximization criterion, C1 ($j = 1$), maximum of its values (i.e., 0.185, 0.317, 0.555, 0.731 and 0.948) in the given ACM is 0.948. Hence, the normalized value of C1 for the second alternative is $\frac{0.317}{0.948} = 0.3344$. Normalized values of C1 for all five alternatives are shown in the second column of Table 4.2.

On the other hand, for the minimization criterion, C2 ($j = 2$), minimum of its values (i.e., 2.33, 1.08, 6.45, 8.88 and 7.33) in the given ACM is 1.08. Hence, the normalized value of C2 for the first alternative is $\frac{1.08}{2.33} = 0.4635$. Normalized values of C2 for all alternatives are given in the third column of Table 4.2, whereas normalized values of C3 are in the fourth column.

TABLE 4.2

Values of ACM in Table 4.1 After Max Normalization.

Alternatives	C1	C2	C3
A1	0.1951	0.4635	0.5347
A2	0.3344	1.0000	0.3510
A3	0.5854	0.1674	0.2049
A4	0.7711	0.1216	1.0000
A5	1.0000	0.1461	0.6090

4.3 Max–Min Normalization

In the max–min normalization, values of a benefit criterion for all m alternatives are normalized by:

$$F_{ij} = \frac{f_{ij} - \min\limits_{k \in \{1,2,\ldots,m\}} f_{kj}}{\max\limits_{k \in \{1,2,\ldots,m\}} f_{kj} - \min\limits_{k \in \{1,2,\ldots,m\}} f_{kj}} \tag{4.2a}$$

This equation is used for every benefit criterion in ACM. On the other hand, the values of a cost criterion for all m alternatives are normalized by a different equation as follows:

$$F_{ij} = \frac{\max\limits_{k \in \{1,2,\ldots,m\}} f_{kj} - f_{ij}}{\max\limits_{k \in \{1,2,\ldots,m\}} f_{kj} - \min\limits_{k \in \{1,2,\ldots,m\}} f_{kj}} \tag{4.2b}$$

After max–min normalization, the maximum and minimum of any criterion are exactly 1.0 and 0.0, respectively. Moreover, values of a cost criterion become like those of a benefit criterion (i.e., larger values are preferred).

NUMERICAL CALCULATIONS (USING ACM FROM TABLE 4.1)

For instance, in the case of a maximization criterion, C1 ($j = 1$), maximum and minimum of its values (i.e., 0.185, 0.317, 0.555, 0.731 and 0.948) in the given ACM are 0.948 and 0.185, respectively. Hence, the normalized value of C1 for the second alternative is $\frac{0.317-0.185}{0.948-0.185} = 0.1730$. Normalized values of C1 for all five alternatives are in the second column of Table 4.3.

On the other hand, for the minimization criterion, C2 ($j = 2$), maximum and minimum of its values (i.e., 2.33, 1.08, 6.45, 8.88 and 7.33) in the given ACM are 8.88 and 1.08, respectively. Hence, the normalized value of C2 for the first alternative is $\frac{8.88-2.33}{8.88-1.08} = 0.8397$. Normalized values of C2 for all alternatives are given in the third column of Table 4.3, whereas normalized values of C3 are in the fourth column.

TABLE 4.3

Values of ACM in Table 4.1 After Max–Min Normalization.

Alternatives	C1	C2	C3
A1	0.0000	0.8397	0.4148
A2	0.1730	1.0000	0.1837
A3	0.4849	0.3115	0.0000
A4	0.7156	0.0000	1.0000
A5	1.0000	0.1910	0.5081

4.4 Sum Normalization

In sum normalization, each value of a criterion (say, j) is divided by the sum of values of the criterion j for all m alternatives:

$$F_{ij} = \frac{f_{ij}}{\sum_{k=1}^{m} f_{kj}} \tag{4.3a}$$

This equation can be used for both benefit and cost criteria (i.e., for all criteria in ACM), which is followed in the complex proportional assessment (COPRAS) method described in Chapter 8. After sum normalization using Eq. 4.3a, values of any (either benefit or cost) criterion will be below 1.0 and some of them can be negative if the corresponding original values in ACM are negative. In a rare case, the sum in the denominator of Eq. 4.3a can be zero; then, sum normalization cannot be employed.

Instead of Eq. 4.3a, the following inverse equation is available for normalizing the values of a cost criterion by sum normalization (Kosareva et al., 2018).

$$F_{ij} = \frac{\left(\frac{1}{f_{ij}}\right)}{\sum_{k=1}^{m}\left(\frac{1}{f_{kj}}\right)} \tag{4.3b}$$

This can be interpreted as transforming a cost criterion by taking its reciprocal to a benefit criterion and then normalization as in Eq. 4.3a. So, the normalized values of the cost criterion become like those of a benefit criterion (i.e., larger values are preferred). Assuming all values of a cost criterion are positive, the normalized values by the inverse equation (Eq. 4.3b) will be less than 1.0 and more than 0.0. There can be problems if one or more values of a cost criterion are negative, while all others are positive.

NUMERICAL CALCULATIONS (USING ACM FROM TABLE 4.1)

For instance, consider the maximization criterion, C3 ($j = 3$); sum of its values for all five alternatives (i.e., 454, 298, 174, 849 and 517) is 2292. Then, normalized C3 for alternative A1 is $\frac{454}{2292} = 0.1981$. Normalized values of C3 for all five alternatives are given in the last column of Table 4.4, wherein the second column has normalized values of C1 by Eq. 4.3a.

For the minimization criterion, C2 ($j = 2$), let's first use Eq. 4.3a for normalization. These calculations are like those for the C3 criterion. The resulting normalized values for C2 are in the third column of Table 4.4. Among the five alternatives, A2 continues to have the lowest value of C2, as in the original ACM (Table 4.1). In other words, normalization by Eq. 4.3a does not change a cost criterion to a benefit criterion.

Now, consider Eq. 4.3b for normalization of C2 values. Reciprocals of C2 values in ACM (Table 4.1) for the five alternatives are, respectively, $\frac{1}{2.33} = 0.4292$, $\frac{1}{1.08} = 0.9259$, $\frac{1}{6.45} = 0.1550$, $\frac{1}{8.88} = 0.1126$ and $\frac{1}{7.39} = 0.1353$ and their sum is 1.7581. Then, the normalized value of C2 for alternative A1 by Eq. 4.3b is $\frac{0.4292}{1.7581} = 0.2441$. This and the normalized values of C2 for the remaining four alternatives are in the fourth column of Table 4.4. In this column, the maximum of normalized C2 values is that of alternative A2. In other words, Eq. 4.3b changes a cost criterion to a benefit criterion.

TABLE 4.4

Values of ACM in Table 4.1 After Sum Normalization.

Alternatives	C1	C2 (by Eq. 4.3a)	C2 (by Eq. 4.3b)	C3
A1	0.0676	0.0892	0.2441	0.1981
A2	0.1159	0.0413	0.5267	0.1300
A3	0.2029	0.2468	0.0882	0.0759
A4	0.2672	0.3398	0.0641	0.3704
A5	0.3465	0.2828	0.0770	0.2256

4.5 Vector Normalization

In vector normalization, each value of a criterion (say, j) is divided by the square root of the sum of squares of values of the criterion j for all m alternatives:

$$F_{ij} = \frac{f_{ij}}{\sqrt{\sum_{k=1}^{m} f_{kj}^{2}}} \tag{4.4a}$$

This equation can be used for both benefit and cost criteria in ACM, which is utilized in the technique for order of preference by similarity to ideal solution (TOPSIS) and preference ranking on the basis of ideal-average distance (PROBID) methods (Wang et al., 2021), as described in Chapter 7. After vector normalization using Eq. 4.4a, values of any (either benefit or cost) criterion will be below 1.0 and some of them can be negative if the corresponding original values in ACM are negative.

Instead of Eq. 4.4a, the following inverse equation is available for normalizing the values of a cost criterion by vector normalization (Kosareva et al., 2018).

$$F_{ij} = 1 - \frac{f_{ij}}{\sqrt{\sum_{k=1}^{m} f_{kj}^{2}}} \tag{4.4b}$$

This is equivalent to transforming a cost criterion to a benefit criterion through subtracting the normalized value by Eq. 4.4a from 1.0. So, the normalized values of the cost criterion become like those of a benefit criterion (i.e., larger values are preferred). Assuming all values of a cost criterion are positive, the normalized values by the inverse equation (Eq. 4.4b) will be less than 1.0 and more than 0.0.

NUMERICAL CALCULATIONS (USING ACM FROM TABLE 4.1)

For instance, consider the maximization criterion, C3 $(j = 3)$; sum of squares of its values for all five alternatives is $454^2 + 298^2 + 174^2 + 849^2 + 517^2 = 1{,}313{,}286$. Then, the denominator in Eq. 4.4a is $\sqrt{1{,}313{,}286} = 1{,}146$. Then, normalized C3 for alternative A1 is $\frac{454}{1146} = 0.3962$. Normalized values of C3 for all five alternatives are given in the last column of Table 4.5, wherein the second column has normalized values of C1 by Eq. 4.4a.

For the minimization criterion, C2 ($j = 2$), let's first use Eq. 4.4a for normalization. These calculations are like those for the C3 criterion. The resulting normalized values for C2 are in the third column of Table 4.4. Among the five alternatives, A2 continues to have the lowest value of C2, as in the original ACM (Table 4.1). In other words, normalization by Eq. 4.4a does not change a cost criterion to a benefit criterion.

Now, consider Eq. 4.4b for normalization of C2 values. Then, the normalized value of C2 for alternative A1 by Eq. 4.4b is 1 – 0.1729 = 0.8271. This and the normalized values of C2 for the remaining four alternatives are in the fourth column of Table 4.4. In this column, the maximum of normalized C2 values is that of alternative A2. In other words, Eq. 4.3b changes a cost criterion to a benefit criterion.

TABLE 4.5

Values of ACM in Table 4.1 After Vector Normalization.

Alternatives	C1	C2 (by Eq. 4.4a)	C2 (by Eq. 4.4b)	C3
A1	0.1351	0.1729	0.8271	0.3962
A2	0.2315	0.0801	0.9199	0.2600
A3	0.4052	0.4785	0.5215	0.1518
A4	0.5337	0.6588	0.3412	0.7408
A5	0.6922	0.5483	0.4517	0.4511

4.6 Discussion

Normalized values (in Tables 4.2 to 4.5) by max, max–min, sum and vector methods for the maximization/benefit criterion, C3 in ACM data in Table 4.1, are compared in Figure 4.1. Clearly, normalized values by these methods are linear with respect to the original values of C3 and are within 0–1, which is mainly due to all positive original values for C3. As expected, the range of normalized values by the max–min and sum methods is, respectively, wide (0–1) and narrow (about 0.3) in this example. The range of normalized values by the max and vector methods is in between (about 0.8 and 0.6, respectively).

Next, normalized values by the four methods for the minimization/cost criterion, C2, are compared in Figure 4.2. Recall that sum and vector normalization methods have two alternatives (Eqs. 4.3a and 4.3b and Eqs. 4.4a and 4.4b) for the minimization criterion. Hence, Figure 4.2 has two curves for each of the sum and vector methods. In this plot, some trends such as ranges and linearity are like those in Figure 4.1. However, there are some differences

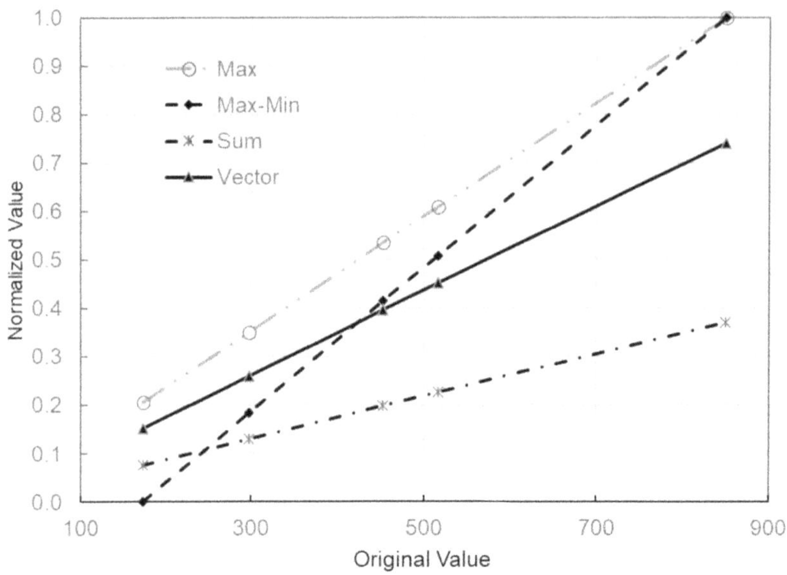

FIGURE 4.1
Plot of normalized values by max, max–min, sum and vector normalization methods versus original value: benefit criterion, C3, to be maximized.

too. First, max normalization is not linear (due to the reciprocal relationship with the original value in Eq. 4.1b). This nonlinearity can also be seen after sum normalization via Eq. 4.3b, which involves reciprocals. Second, after the max–min method, the minimum original value of C2 corresponds to the maximum (i.e., 1.0) and the maximum original value corresponds to the minimum (i.e., 0.0). In other words, the C2 criterion, after normalization, becomes like a benefit criterion, as noted earlier. This directional change occurs after max normalization, sum normalization via Eq. 4.3b and vector normalization via Eq. 4.4b, as well.

For generalizing the range of normalized values by the four methods, assume values of a criterion for all alternatives are greater than 0 and two cases: (1) The range of values of a criterion is narrow, for example, within 5% and (2) a wider range of criterion values (say, the ratio of maximum to minimum of the criterion is λ). Table 4.6 summarizes the range of normalized values for the combination of these two cases and four normalization methods, as an expansion of that described in Wang et al. (2020). This table and Figures 4.1 and 4.2 provide a deeper understanding of the four normalization methods and the resulting normalized values, which will be helpful in choosing an appropriate method.

Besides the four normalization methods covered in this chapter, there are other methods such as the logarithmic transformation of Zavadskas and

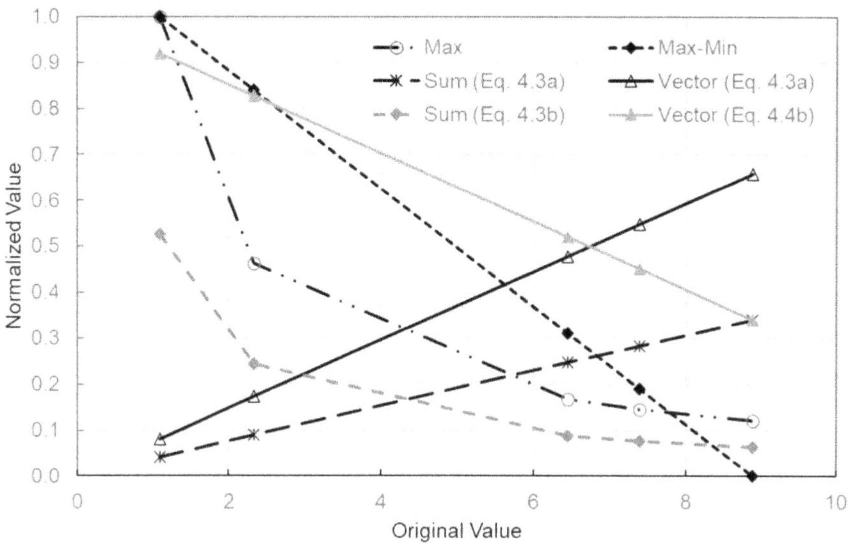

FIGURE 4.2
Plot of normalized values by max, max–min, sum and vector normalization methods versus original value: cost criterion, C2, to be minimized.

TABLE 4.6

Range of Normalized Values by Max, Max–Min, Sum and Vector Methods in Two Cases of Narrow and Wider Ranges of Criterion Values; m is the Number of Alternatives.

Normalization Method	Case (i): Narrow Range of Criterion Values	Case (ii): Wider Range of Criterion Values by a Factor of λ
Max	All less than 1.0 and near 1.0	$\frac{1}{\lambda}$ to 1.0
Max–min	0.0 to 1.0	0.0 to 1.0
Sum (Eq. 4.3a or 4.3b)	All near $1/m$	$\frac{1}{1+\lambda(m-1)}$ to $\frac{\lambda}{\lambda+(m-1)}$
Vector (Eq. 4.4a or 4.4b)	All near $1/\sqrt{m}$	$\frac{1}{\sqrt{1+\lambda^2(m-1)}}$ to $\frac{\lambda}{\sqrt{\lambda^2+(m-1)}}$

Turskis (2008). Kosareva et al. (2018) and Wang et al. (2024) present equations of normalization methods for both benefit and cost criteria. Aytekin (2021) lists and analyzes many normalization methods available in the literature. Further, Aytekin (2021) recommends the most compatible normalization method(s) for use in ranking methods such as gray/grey relational analysis (GRA), TOPSIS and VIKOR, which are described in Chapter 7 of this book;

SAW, AHP and MOO on the basis of ratio analysis (MOORA), which are covered in Chapter 8; and ELECTRE and PROMETHEE, which are presented in Chapter 9. In some cases, the recommended method in Aytekin (2021) is different from that in the original reference of the MCDM method. Baydaş et al. (2023) studied the normalization method selection for MCDM. Other studies on normalization methods include Vafaei et al. (2016) in the context of AHP and Vafaei et al. (2021) in the context of TOPSIS.

4.7 Numerical Difficulties

During normalization, there can be numerical difficulties in some situations. Division by zero can occur in the following situations: (1) in max normalization, when the maximum value of a benefit criterion is zero (Eq. 4.1a) and when any value of a cost criterion is zero (Eq. 4.1b); (2) in sum normalization, when the sum is zero, which can occur when a criterion has some positive and some negative values (Eq. 4.3a); and (3) when any value of a cost criterion is zero (Eq. 4.3b). Further, all values of a cost criterion by max normalization become zero if the minimum of that cost criterion is zero. Although these situations may not be common, care should be exercised if the value of a criterion is zero or values of a criterion include both positive and negative values (in which case, some normalized values can be outside the range of 0–1). See the following example and Exercise 4.4.

NUMERICAL CALCULATIONS FOR A MINIMIZATION CRITERION HAVING NEGATIVE AND POSITIVE VALUES

Consider the minimization of a criterion with the values in Table 4.7 for five alternatives. These are created by changing the sign of C2 values for A1 and A5 in ACM in Table 4.1 from positive to negative. Thus, C2 has some negative and some positive values.

TABLE 4.7

Modified Values of Criterion C2 to Be Minimized.

Alternatives	C2 (Min)
A1	−2.33
A2	1.08
A3	6.45
A4	8.88
A5	−7.39

Application of max, max–min, sum and vector normalization methods results in the normalized values for five alternatives presented in Table 4.8. Calculations in this application are like those described in Sections 4.2–4.5 of this chapter; however, care should be exercised to retain the sign of values for A1 and A5 alternatives.

TABLE 4.8

Normalized Values by Max, Max–Min, Sum and Vector Methods for the Minimization Criterion Having Some Negative Values and Some Positive Values.

Alternatives	Max Method	Max–Min Method	Sum (Eq. 4.3a)	Sum (Eq. 4.3b)	Vector (Eq. 4.4a)	Vector (Eq. 4.4b)
A1	3.1717	0.6890	−0.3483	−0.6822	−0.1729	1.1729
A2	−6.8426	0.4794	0.1614	1.4719	0.0801	0.9199
A3	−1.1457	0.1494	0.9641	0.2465	0.4785	0.5215
A4	−0.8322	0.0000	1.3274	0.1790	0.6588	0.3412
A5	1.0000	1.0000	−1.1046	−0.2151	−0.5483	1.5483

Results in Table 4.8 clearly demonstrate that negative/positive original values of a criterion can affect the sign and range of normalized values, which depend on the normalization method. The max–min method is the least affected with the normalized values remaining in the range of 0–1. In the case of max, sum and vector methods, some normalized values can be negative and the range can be wider (i.e., larger than 0.0–1.0). These can be explained by reference to their respective equations.

Some numerical difficulties in the normalization of ACM data can be avoided by transforming each cost criterion into a benefit criterion by subtracting each value of the cost criterion by a constant, which is (slightly) more than the maximum of the criterion values. This is illustrated in Table 4.9 for the criterion values in Table 4.7; here, the constant of 9.0 is assumed; however, it can be any value more than 8.88. After this transformation, criterion C2 can be treated as a benefit criterion, which simplifies normalization by any of the four methods presented in this chapter. The alternative A5 originally has the minimum value of −7.39 for the criterion; after transformation, it has the maximum value of 16.39. The transformed values in the last column of Table 4.9 can then be normalized by any of the methods including the four described in this chapter.

TABLE 4.9

Original Values of Cost Criterion C2 and Their Transformed Values for Maximization (i.e., Benefit Criterion).

Alternatives	Original Values to Be Minimized	Transformed Values to Be Maximized
A1	−2.33	9.0 – (−2.33) = 11.33
A2	1.08	9.0 – 1.08 = 7.92
A3	6.45	9.0 – 6.45 = 2.55
A4	8.88	9.0 – 8.88 = 0.12
A5	−7.39	9.0 – (−7.39) = 16.39

4.8 Summary

Main points in this chapter on normalization methods can be summarized as follows:

1. In the application of MCDM/ranking methods, normalization of criterion values is generally required because of different magnitudes and dimensions (units) of each criterion.
2. Of the methods reported in the literature, this chapter describes max, max–min, sum and vector normalization methods. In these methods, normalization equations for benefit and cost criteria may be different.
3. There is an alternative equation for normalizing a cost criterion by the sum and vector methods. Hence, carefully check and use normalization equations, for example, for reproducing (reported) results by an MCDM method.
4. Potential computational difficulties in the use of normalization methods are highlighted; these are due to division by zero and a mix of negative/positive values of a criterion. Some of these can be avoided by transforming a cost criterion into a benefit criterion by subtracting its values from a constant, which is a sufficiently large positive value.

References

Aytekin, A. (2021). Comparative analysis of normalization techniques in the context of MCDM problems. *Decision Making: Applications in Management and Engineering*, *4*(2), 1–25.

Baydaş, M., Eren, T., & İyibildiren, M. (2023). Normalization technique selection for MCDM methods: A flexible and conjunctural solution that can adapt to changes in financial data types. *Necmettin Erbakan Üniversitesi Siyasal Bilgiler Fakültesi Dergisi*, *5*(Özel Sayı), 148–164.

Kosareva, N., Krylovas, A., & Zavadskas, E. K. (2018). Statistical analysis of MCDM data normalization methods using Monte Carlo approach: The case of ternary estimates matrix. *Economic Computation and Economic Cybernetics Studies and Research*, *52*(4), 159–175.

Nabavi, S. R., Wang, Z., & Rodríguez, M. L. (2024). Multi-objective optimization and multi-criteria decision-making approach to design a multi-tubular packed-bed membrane reactor in oxidative dehydrogenation of ethane. *Energy & Fuels*, *39*(1), 491–503.

Vafaei, N., Ribeiro, R. A., & Camarinha-Matos, L. M. (2016, April). Normalization techniques for multi-criteria decision making: Analytical hierarchy process case study. In *7th Doctoral Conference on Computing, Electrical and Industrial Systems (DoCEIS)* (pp. 261–269). Costa de Caparica, Portugal. https://doi.org/10.1007/978-3-319-31165-4_26. hal-01438251

Vafaei, N., Ribeiro, R. A., & Camarinha-Matos, L. M. (2021, July). Assessing normalization techniques for TOPSIS method. In *12th Doctoral Conference on Computing, Electrical and Industrial Systems (DoCEIS)* (pp. 132–141). Costa de Caparica, Portugal. https://doi.org/10.1007/978-3-030-78288-7_13. hal-03685943

Wang, Z., Baydaş, M., Stević, Ž., Özçil, A., Irfan, S. A., Wu, Z., & Rangaiah, G. P. (2023). Comparison of fuzzy and crisp decision matrices: An evaluation on PROBID and sPROBID multi-criteria decision-making methods. *Demonstratio Mathematica*, *56*(1), 20230117.

Wang, Z., Li, J., Rangaiah, G. P., & Wu, Z. (2022). Machine learning aided multi-objective optimization and multi-criteria decision making: Framework and two applications in chemical engineering. *Computers & Chemical Engineering*, *165*, 107945.

Wang, Z., Nabavi, S. R., & Rangaiah, G. P. (2024). Multi-criteria decision making in chemical and process engineering: Methods, progress and potential. *Processes*, *12*(11), 2532.

Wang, Z., Parhi, S. S., Rangaiah, G. P., & Jana, A. K. (2020). Analysis of weighting and selection methods for Pareto-optimal solutions of multi-objective optimization in chemical engineering applications. *Industrial & Engineering Chemistry Research*, *59*(33), 14850–14867.

Wang, Z., Rangaiah, G. P., & Wang, X. (2021). Preference ranking on the basis of ideal-average distance method for multi-criteria decision-making. *Industrial & Engineering Chemistry Research*, *60*(30), 11216–11230.

Zavadskas, E. K., & Turskis, Z. (2008). A new logarithmic normalization method in games theory. *Informatica*, *19*(2), 303–314.

Exercises

E4.1 Normalize ACM data for three criteria and four alternatives in the following table, using (1) max, (2) max–min, (3) sum and (4) vector methods. Compare the ranges of the normalized values by these methods.

Alternatives	C1 (Max)	C2 (Max)	C3 (Min)
A1	0.93	600	8.25
A2	0.51	700	6.33
A3	0.77	500	3.16
A4	0.82	400	2.98

E4.2 Normalize ACM data for four criteria and six alternatives in the following table, using (1) max, (2) max–min, (3) sum and (4) vector methods. Compare the ranges of the normalized values by these methods.

Alternatives	C1 (Max)	C2 (Min)	C3 (Min)	C4 (Max)
A1	119	24.7	77.4	8
A2	247	27.8	52.0	7
A3	235	15.7	77.0	3
A4	132	15.1	67.2	7
A5	190	26.7	78.5	10
A6	184	18.7	60.3	6

E4.3 Normalize ACM data for two criteria and five alternatives in the following table, using (1) max, (2) max–min, (3) sum and (4) vector methods. Compare the ranges of the normalized values by these methods.

Alternatives	C1 (Min)	C2 (Max)
A1	1.1	135.2
A2	1.7	147.1
A3	1.9	189.3
A4	2.1	226.8
A5	2.4	238.3

E4.4 Normalize the following values of a benefit criterion, C1, for five alternatives using (1) max, (2) max–min, (3) sum and (4) vector methods. Comment on the normalized values (e.g., negative or positive values and range).

Alternatives	C1 (Max)
A1	−0.185
A2	−0.317
A3	0.555
A4	0.731
A5	0.948

E4.5 Answer each of the following questions: (1) Name a normalization method, which gives normalized values of a criterion covering the entire range of 0.0–1.0, (2) name a normalization method, which gives normalized values of a criterion in the smallest range and (3) name a normalization method, which gives normalized values in the same range even if there are some negative values of a criterion.

5

Objective Weighting Methods

5.1 Overview

In multi-ctieria decision-making (MCDM), determining the weight value for each criterion is crucial. Although decision-makers can give weights or preferences themselves, they may encounter difficulties in doing so and/or there can be biases in the resulting weights. In such instances, objective weighting methods are preferred. Here, "objective" does not mean criterion; instead, it signifies that the methods are free from subjective personal opinions. Almost all objective weighting methods require the criteria values of all alternatives (i.e., alternatives-criteria matrix, ACM data) as well as normalization (covered in the previous chapter) of ACM data before weight computations. Another category of methods, namely, subjective weighting methods that require inputs/preferences from decision-makers (Wang et al., 2024), is covered in the next chapter (Chapter 6).

The present chapter delves into seven objective weighting methods, detailing their principles and equations. Many of these methods utilize the values of the criteria for the alternatives in ACM to determine the weights of the criteria. Hence, the calculated weight of a criterion by an objective weighting method depends on the criterion values (and their range) for the alternatives. When all alternatives in ACM have similar values (i.e., a very narrow range) for a criterion, then that criterion is not helpful in choosing an alternative and so it should be given a lower weight. Conversely, a criterion with more diverse values (i.e., a wider range) across the alternatives contributes more to differentiating between alternatives and should accordingly receive a higher weight. This is loosely like the logic in the principal component analysis (PCA) technique, where higher variance/spread is regarded as carrying more information.

Each of the seven methods covered in this chapter is elucidated with a simple numerical example, ensuring readers gain a clear understanding of their inner workings. Although different normalization methods can be used before computing criteria weights, each objective weighting method is described and illustrated using the normalization method as in the original/

DOI: 10.1201/9781003530435-5

cited reference of that method. The remaining sections in this chapter are as follows:

Section 5.2 Mean Weighting (MW) Method
Section 5.3 Standard Deviation Weighting (SDW) Method
Section 5.4 Statistical Variance Weighting (SVW) Method
Section 5.5 Entropy Weighting (EW) Method
Section 5.6 Criteria Importance Through Intercriteria Correlation (CRITIC) Method
Section 5.7 Gini Coefficient Weighting (GCW) Method
Section 5.8 Method Based on the Removal Effects of Criteria (MEREC)
Section 5.9 Discussion
Section 5.10 Summary

In the above, the first three methods are simple and they are presented in alphabetical order. Then, the next four methods are described in chronological order. The MEREC is very recent. Each of the EW, CRITIC, GCW and MEREC methods brings in a new perspective for calculating criteria weights.

The learning outcomes of this chapter are to (1) describe the objective weighting methods covered in this chapter; (2) state similarities, differences and relative merits of these methods; and (3) select and apply any of them to the given ACM data.

Unless otherwise stated, the notation used throughout this chapter is as follows:

- $i \in \{1,2,\dots,m\}$ and $j \in \{1,2,\dots,n\}$, where m is the number of rows (i.e., alternatives) and n is the number of columns (i.e., criteria) in the ACM.
- f_{ij} is the value of the jth criterion for the ith alternative in the original ACM.
- F_{ij} is the normalized value of f_{ij}.
- G_j is the Gini coefficient of values of the jth criterion.
- P_i is the overall performance of the ith alternative considering all criteria in ACM
- P_{ij} is the overall performance of the ith alternative after removal of the jth criterion in ACM.
- R_j is the overall effect of removing j^{th} criterion in ACM.
- $r_{jk} = r_{kj}$ is the correlation coefficient between the jth and k^{th} criteria.
- w_j is the weight assigned to the jth criterion and $\sum_{j=1}^{n} w_j = 1$.

TABLE 5.1

ACM Employed for Illustration of Objective Weighting Methods (C1 and C3 for Maximization; C2 for Minimization).

Alternatives	C1 (Max)	C2 (Min)	C3 (Max)
A1	0.185	2.33	454
A2	0.317	1.08	298
A3	0.555	6.45	174
A4	0.731	8.88	849
A5	0.948	7.39	517

- σ_j is the standard deviation (SD) of the values of the jth criterion.
- σ_j^2 is the variance of values of the jth criterion.

Additionally, the ACM data in Table 5.1 are employed to illustrate the objective weighting methods covered in this chapter. This ACM is made up of five alternatives (i.e., $m = 5$ and $i \in \{1,2,3,4,5\}$) and three criteria (i.e., $n = 3$ and $j \in \{1,2,3\}$). Out of the three criteria (C1, C2 and C3), C1 and C3 are benefit criteria to be maximized, while C2 is a cost criterion to be minimized. All five alternatives (A1, A2, A3, A4 and A5) are non-dominated, which means that, among the alternatives, the improvement in one criterion will inevitably result in degradation in at least one other criterion (i.e., a trade-off among the criteria).

5.2 Mean Weighting (MW) Method

The MW method is the simplest of all weighting methods. It assumes equal importance for all criteria in an MCDM application and so the same weight is taken for every criterion in ACM. Further, in this and all weighting methods, the sum of weights of all criteria is assumed to be equal to unity. Both these assumptions imply that $w_j = \frac{1}{n}$ for $j = 1,2,\ldots,n$ by the MW (or equal or same weighting) method. Thus, this method requires just the value of n in an MCDM application and not the complete ACM data (i.e., number of alternatives and values of criteria). For example, for ACM in Table 5.1, only $n = 3$ is required and criteria weights are $w_1 = w_2 = w_3 = \frac{1}{3} = 0.3333$. Obviously, the MW method is appropriate if the decision-makers consider that all criteria in their application have the same importance and so the same weight.

TABLE 5.2

Normalized Values of ACM in Table 5.1 After Max–Min Normalization.

Alternatives	C1	C2	C3
A1	0.0000	0.8397	0.4148
A2	0.1730	1.0000	0.1837
A3	0.4849	0.3115	0.0000
A4	0.7156	0.0000	1.0000
A5	1.0000	0.1910	0.5081

5.3 Standard Deviation Weighting (SDW) Method

As the name implies, the criteria weights by the SDW method are based on the statistical quantity, SD, σ of a set of values. Steps in this method are as follows:

Step 1. Normalize ACM data using a normalization method. This can be performed by the max–min method, as in Diakoulaki et al. (1995), although any other normalization method (including those covered in Chapter 4) can be chosen. The equations for the max–min method of normalization and their illustration are given in Section 4.3. After max–min normalization of ACM data in Table 5.1, normalized values (F_{ij}) of the three criteria are presented in Table 5.2, which is the same as Table 4.3. After max–min normalization, values of a cost criterion become like those of a benefit criterion (i.e., larger values are preferred). Hence, both benefit and cost criteria can then be treated in the same way. Further, the values of each criterion are in the closed interval of 0.0–1.0.

Step 2. Compute the SD of normalized values (F_{ij}) of each criterion by the usual definition:

$$\sigma_j = \sqrt{\frac{\sum_{i=1}^{m}\left(F_{ij} - \bar{F}_j\right)^2}{m}} \quad \text{for } j \in (1,2,\ldots,n) \tag{5.1}$$

Here, $\bar{F}_j = \frac{1}{m}\sum_{i=1}^{m} F_{ij}$ is the arithmetic mean of the normalized values of the jth criterion.

NUMERICAL CALCULATIONS (USING ACM FROM TABLE 5.1)

Consider normalized values of C2 in Table 5.2, where $m = 5$. In this case, $j = 2$ and F_{i2} values are 0.8397, 1.0000, 0.3115, 0.0000 and 0.1910 and their mean is $\bar{F}_2 = \frac{1}{5}(0.8397 + 1.0 + 0.3115 + 0.0 + 0.1910) = 0.4684$. Then, using Eq. 5.1,

$$\sigma_2 = \sqrt{\frac{(0.8397-0.4684)^2+(1.0-0.4684)^2+(0.3115-0.4684)^2+(0.0-0.4684)^2+(0.1910-0.4684)^2}{5}} = 0.3851$$

Similar calculations for criteria, C1 and C3, give $\sigma_1 = 0.3607$ and $\sigma_3 = 0.3395$. These SDs are less than unity, since they are based on normalized criteria values.

Step 3. Finally, calculate the weight of each criterion by:

$$w_j = \frac{\sigma_j}{\sum_{k=1}^{n}\sigma_k} \quad \text{for } j \in (1,2,\ldots,n) \tag{5.2}$$

Thus, weights are found by "normalizing" SD values of all criteria in the MCDM application and so they are proportional to SD values. Criteria weights by SDW and many objective weighting methods depend on the normalization method in Step 1.

NUMERICAL CALCULATIONS (USING ACM FROM TABLE 5.1)

From Step 2, $\sigma_1 = 0.3607$, $\sigma_2 = 0.3851$ and $\sigma_3 = 0.3395$, whose sum is 1.0853. The weights of C1, C2 and C3 are, respectively, $w_1 = \frac{0.3607}{1.0853} = 0.3323$, $w_2 = \frac{0.3851}{1.0853} = 0.3548$ and $w_3 = \frac{0.3395}{1.0853} = 0.3129$. Clearly, these weights are comparable since the SDs of the normalized values of the three criteria are comparable.

5.4 Statistical Variance Weighting (SVW) Method

The SVW method is based on another common statistic, namely, variance, which is equal to the square of SD. Steps in this method are as follows:

Step 1. Normalize ACM data using a normalization method. For simplicity and comparison, let's use max–min normalization employed in the SDW method, although any other method can be chosen. For example, Rao and Patel (2010) used sum normalization described in Section 4.4. Calculations and results of this step of applying max–min normalization are the same as those in Step 1 of the SDW method (e.g., Table 5.2).

Step 2. Compute the variance (σ_j^2) of the normalized values (F_{ij}) of each criterion by the usual definition:

$$\sigma_j^2 = \frac{\sum_{i=1}^{m}\left(F_{ij} - \bar{F}_j\right)^2}{m} \text{ for } j \in (1,2,\ldots,n) \tag{5.3}$$

NUMERICAL CALCULATIONS (USING ACM FROM TABLE 5.1)

Consider normalized values of C2 in Table 5.2, where $m = 5$. In this case, $j = 2$ and F_{i2} values are 0.8397, 1.0000, 0.3115, 0.0000 and 0.1910 and their mean is $\bar{F}_2 = \frac{1}{5}(0.8397 + 1.0 + 0.3115 + 0.0 + 0.1910) = 0.4684$. Then, using Eq. 5.3,

$$\sigma_2^2 = \frac{(0.8397 - 0.4684)^2 + (1.0 - 0.4684)^2 + (0.3115 - 0.4684)^2 + (0.0 - 0.4684)^2 + (0.1910 - 0.4684)^2}{5} = 0.1483$$

Similar calculations for criteria, C1 and C3, give $\sigma_1^2 = 0.1301$ and $\sigma_3^2 = 01153$. These variances are less than unity, since they are based on normalized criteria values. Also, due to squaring, variance values differ relatively more than SD values.

Step 3. Finally, calculate the weight of each criterion by:

$$w_j = \frac{\sigma_j^2}{\sum_{k=1}^{n} \sigma_k^2} \quad \text{for} \quad j \in (1,2,\ldots,n \tag{5.4}$$

Thus, like in the SDW method, weights are found by "normalizing" variance values of all criteria in the MCDM application and so they are proportional to variances. Criteria weights by SVW and many objective weighting methods depend on the normalization method in Step 1.

NUMERICAL CALCULATIONS (USING ACM FROM TABLE 5.1)

Calculations in Step 2 gave $\sigma_1^2 = 0.1301$, $\sigma_2^2 = 0.1483$ and $\sigma_3^2 = 0.1153$, whose sum is 0.3937. The weights of C1, C2 and C3 are, respectively,

$w_1 = \frac{0.1301}{0.3937} = 0.3305, w_2 = \frac{0.1483}{0.3937} = 0.3767$ and $w_3 = \frac{0.1153}{0.3937} = 0.2929$.
These weights are comparable since the variances of the normalized values of the three criteria are comparable. Also, they are like those by the SDW method.

5.5 Entropy Weighting (EW) Method

Entropy, a measure of randomness or uncertainty, is an important concept in physical sciences and engineering. The EW method, based on it, can be traced to Hwang and Yoon (1981). It is frequently applied in MCDM applications (Wang et al., 2023). Steps and equations in the EW method are described below.

Step 1. Normalize ACM data using sum normalization (described in Section 4.4). The equation for this is:

$$F_{ij} = \frac{f_{ij}}{\sum_{k=1}^{m} f_{kj}} \tag{5.5}$$

This equation is used for both benefit and cost criteria, as in Hwang and Yoon (1981), who refer F_{ij} as the project outcomes of criterion *j* (and not sum normalization). Some researchers (e.g., Mukhametzyanov, 2021) perform max–min normalization of ACM data and then use the resulting normalized values in Eq. 5.5.

After sum normalization of ACM data in Table 5.1, normalized values (F_{ij}) of the three criteria are presented in Table 5.3. See Section 4.4 for these normalization calculations. Clearly, normalized values after sum normalization are much less than unity.

Step 2. Based on normalized values (Table 5.3), calculate the entropy of the criterion j (E_j) by:

$$E_j = -\frac{1}{\ln(m)} \sum_{i=1}^{m} \left[F_{ij} \times \ln\left(F_{ij}\right) \right] \quad \text{for } j \in (1,2,\ldots,n) \tag{5.6}$$

Here, ln stands for natural logarithm to the base e = 2.71828 and the summation is for all m alternatives. Since the normalized criteria values (F_{ij}) are less than 1.0, $\ln\left(F_{ij}\right)$ and consequently each term in the summation will be less than zero. The negative sign on the right side of the above equation makes E_j positive. If F_{ij} is zero, then its logarithm is infinity, but $F_{ij} \times \ln\left(F_{ij}\right)$ is zero. Further, $\ln\left(F_{ij}\right)$ is not

TABLE 5.3

Normalized Values of ACM in Table 4.1 After Sum Normalization.

Alternatives	C1	C2	C3
A1	0.0676	0.0892	0.1981
A2	0.1159	0.0413	0.1300
A3	0.2029	0.2468	0.0759
A4	0.2672	0.3398	0.3704
A5	0.3465	0.2828	0.2256

defined for negative values of F_{ij}. These situations pose numerical difficulties in calculating the entropy by Eq. 5.6.

Essentially, the EW method is based on the probabilistic approach to quantify informational uncertainty. Specifically, E_j by Eq. 5.6 is the Shannon entropy of uncertainty of the probability distribution of a set of discrete values (i.e., F_{ij} for i = 1, 2, . . ., m). Let's discuss its values. As F_{ij} increases from 0.0 to 1.0, the term $F_{ij} \times \ln(F_{ij})$ in the summation in Eq. 5.6 decreases from 0.0 to a minimum of -0.368 (at $F_{ij} = 0.368 = 1/\mathrm{e}$) and then increases to 0.0 (at $F_{ij} = 1.0$). Thus, this term is nonlinear and consequently, E_j value depends nonlinearly on the values of F_{ij} for i = 1, 2, . . ., m.

NUMERICAL CALCULATIONS (USING ACM FROM TABLE 5.1)

Consider the normalized values of C2 by sum normalization (given in Table 5.3), where $m = 5$. In this case, $j = 2$ and F_{i2} values are 0.0892, 0.0413, 0.2468, 0.3398 and 0.2828. Using Eq. 5.6 for these values,

$$E_2 = -\frac{1}{\ln(5)} \begin{bmatrix} 0.0892 \times \ln(0.0892) + 0.0413 \times \ln(0.0413) \\ +0.2468 \times \ln(0.2468) + 0.3398 \times \ln(0.3398) \\ +0.2828 \times \ln(0.2828 \end{bmatrix} = 0.88014$$

Likewise, $E_1 = 0.9167$ and $E_3 = 0.9230$. Here, the entropies of all three criteria are less than 1.0.

Step 3. As the entropy of a criterion, E_j, increases, the information value of that criterion decreases. Hence, $(1 - E_j)$ is used for calculating the weight of each criterion by:

$$w_j = \frac{1 - E_j}{\sum_{k=1}^{n} (1 - E_k)} \qquad \text{for } j \in (1, 2, \ldots, n) \tag{5.7}$$

Thus, like in SDW and SVW methods, weights are found by "normalizing" $(1-E_j)$ values of all criteria in the MCDM application and so they are proportional to $(1-E_j)$. Like the SDW and SVW methods, the criteria weights by the EW method also depend on the normalization method in Step 1.

NUMERICAL CALCULATIONS (USING ACM FROM TABLE 5.1)

From the calculations in Step 2, $1-E_1=1.0-0.9167=0.0833$, $1-E_2=1.0-0.8801=0.1199$ and $1-E_3=1.0-0.9230=0.0770$, whose sum is 0.2802. The weights of C1, C2 and C3 are, respectively,
$w_1=\frac{0.0833}{0.2802}=0.2973, w_2=\frac{0.1199}{0.2802}=0.4279$ and $w_3=\frac{0.0770}{0.2802}=0.2748$.
Not surprisingly, these weights are different from those by the SDW and SVW methods.

5.6 Criteria Importance Through Intercriteria Correlation (CRITIC) Method

Diakoulaki et al. (1995) proposed the CRITIC method; it is a modification of the SDW method to incorporate correlation among the criteria in ACM data. Thus, the CRITIC method integrates two concepts (Nabavi et al., 2023): contrast or variation of values of a criterion (via SD) and conflict among the criteria (via correlation). Diakoulaki et al. (1995) claimed that it generates more balanced weights for a better resolution of trade-offs among the criteria. The steps in the CRITIC method are as follows:

Step 1. Normalize ACM data using the max–min method, as in Diakoulaki et al. (1995). This step is the same as Step 1 in the SDW method described earlier in Section 5.3 and so normalized values will be as those in Table 5.2.

Step 2. Compute the SD of normalized values (F_{ij}) of each criterion. This step is the same as Step 2 in the SDW method. As given in Section 5.3, SDs for normalized ACM data in Table 5.2 are $\sigma_1=0.3607$, $\sigma_2=0.3851$ and $\sigma_3=0.3395$. A larger σ_j suggests that the jth criterion demonstrates greater variability among the alternatives; this further indicates that its values fluctuate considerably and may exert a stronger impact on decision-making (Wang et al., 2025).

Step 3. Calculate the correlation coefficient, r_{jk}, between two criteria, *j* and *k*, by:

$$r_{jk} = r_{kj} = \frac{\sum_{i=1}^{m}\left(F_{ij} - \bar{F}_j\right)\left(F_{ik} - \bar{F}_k\right)}{\sqrt{\sum_{i=1}^{m}\left(F_{ij} - \bar{F}_j\right)^2}\sqrt{\sum_{i=1}^{m}\left(F_{ik} - \bar{F}_k\right)^2}} \quad \text{for } j,k \in (1,2,\ldots,n) \tag{5.8}$$

Here, $\bar{F}_j = \frac{1}{m}\sum_{i=1}^{m} F_{ij}$ and $\bar{F}_k = \frac{1}{m}\sum_{i=1}^{m} F_{ik}$ are the arithmetic means of normalized values of the j th and k th criteria, respectively, for all m alternatives. The correlation coefficient, r_{jk} (Eq. 5.8), simplifies to +1.0 for $j = k$ *and* it is between –1.0 and 1.0 for all $j \neq k$. Values of r_{jk} for $j,k \in (1,2,\ldots,n)$ form a symmetric matrix with elements on the leading diagonal equal to 1.0. A value of 0.0 for r_{jk} indicates that criteria j and k are independent (no association or correlation). And a value of more (less) than 0.0 means criteria j and k are positively (negatively) correlated; that is, values of the two criteria change in the same (opposite) direction.

NUMERICAL CALCULATIONS (USING ACM FROM TABLE 5.1)

For this ACM, normalized values are in Table 5.2, m = 5, n = 3 and $r_{11} = r_{22} = r_{33} = 1.0$. Then, r_{12} $(= r_{21})$, r_{13} $(= r_{31})$ and r_{23} $(= r_{32})$ need to be calculated by Eq. 5.8. For example, r_{12} $(= r_{21})$ can be calculated as follows:

$$\bar{F}_1 = \frac{1}{5}(0.0 + 0.1730 + 0.4849 + 0.7156 + 1.0) = 0.4747$$

$$\bar{F}_2 = \frac{1}{5}(0.8397 + 1.0 + 0.3115 + 0.0 + 0.1910) = 0.4684$$

$$r_{12} = r_{21}$$

$$= \frac{\begin{array}{c}(0.0-0.4747)(0.8397-0.4684)+(0.1730-0.4747)(1.0-0.4684)+\\(0.4849-0.4747)(0.3115-0.4684)+(0.7156-0.4747)(0.0-0.4684)+\\(1.0-0.4747)(0.1910-0.4684)\end{array}}{\sqrt{\begin{array}{c}(0.0-0.4747)^2+(0.1730-0.4747)^2+(0.4849-0.4747)^2+\\(0.7156-0.4747)^2+(1.0-0.4747)^2\end{array}} \times\sqrt{\begin{array}{c}(0.8397-0.4684)^2+(1.0-0.4684)^2+(0.3115-0.4684)^2+\\(0.0-0.4684)^2+(0.1910-0.4684)^2\end{array}}}$$

Clearly, this equation is long; it is better to use a spreadsheet (e.g., Microsoft Excel or the CORREL function in it) for these calculations, which give $r_{12} = r_{21} = -0.8594$.

Likewise, $r_{13} = r_{31} = 0.4172$ and $r_{23} = r_{23} = -0.5473$. The resulting correlation matrix is presented in table form (Table 5.4). Clearly, it is a symmetric matrix.

TABLE 5.4

Correlation Matrix of Normalized Values of ACM in Table 5.1.

	C1	C2	C3
C1	1.0	−0.8594	0.4172
C2	−0.8594	1.0000	−0.5473
C3	0.4172	−0.5473	1.0

Step 4. Combine SD and correlation coefficients to compute the correlation-moderated SD (τ_j) of the jth criterion.

$$\tau_j = \sigma_j \sum_{k=1}^{n} (1 - r_{jk}) \text{ for } j \in (1, 2, \ldots, n) \tag{5.9}$$

To illustrate the effect of r_{jk}, consider a special case of $r_{jk} = 0$ (i.e., no correlation) for all $j \neq k$; then, the correlation matrix is an identity matrix since $r_{jk} = 1.0$ for $j = k$. In this case, $\tau_j = \sigma_j (n - 1)$ for all $j = 1, 2, \ldots, n$; in other words, τ_j is proportional to σ_j for all j. Therefore, weights computed using τ_j (by Eq. 5.10) will be identical to those obtained using the SDW method. Next, consider a single criterion, say, C2, assume $r_{2k} = 1$ (i.e., positively and fully correlated) for all $k = 1, 2, \ldots$, n. Then, $\tau_j = 0$ and consequently, the weight for C2 will be zero. In case correlation coefficients for C2 with all other criteria are negatively (instead of positively) and fully correlated (i.e., $r_{2k} = -1$), then, $\tau_j = 2\sigma_j (n - 1)$ and the weight for C2 (by Eq. 5.10) will be more than that for all other criteria. In summary, a criterion that is overall positively correlated with the others will receive a lower weight, whereas one that is overall negatively correlated will receive a higher weight.

NUMERICAL CALCULATIONS (USING ACM FROM TABLE 5.1)

Recalling n = 3, SDs for normalized ACM data in Table 5.2 are $\sigma_1 = 0.3607$, $\sigma_2 = 0.3851$ and $\sigma_3 = 0.3395$ (found in Step 2). Further, correlation coefficients are in Table 5.4. Then,

$$\tau_1 = \sigma_1 \left[(1 - r_{11}) + (1 - r_{12}) + (1 - r_{13}) \right]$$

$$= 0.3607 \left[(1 - 1) + (1 - (-0.8594)) + \right.$$

$$\left. (1 - 0.4172) \right] = 0.8809$$

Likewise, $\tau_2 = 1.3119$ and $\tau_3 = 0.7232$.

Step 5. Finally, using τ_j values from Step 4, calculate the weight of each criterion by:

$$w_j = \frac{\tau_j}{\sum_{k=1}^{n} \tau_k} \quad \text{for } j \in (1,2,\ldots,n) \tag{5.10}$$

Note that this equation is like Eq. 5.2 in the SDW method except for the use of τ_j (i.e., correlation-modified SD values) instead of σ_j (i.e., SD values). Like the SDW method, weights are found by "normalizing" correlation-moderated SD values of all criteria in the MCDM application and criteria weights by the CRITIC method depend on the normalization method in Step 1.

NUMERICAL CALCULATIONS (USING ACM FROM TABLE 5.1)

Calculations in Step 4 gave: $\tau_1 = 0.8809$, $\tau_2 = 1.3119$ and $\tau_3 = 0.7232$, whose sum is 2.9160. The weights of C1, C2 and C3 are, respectively, $w_1 = \frac{0.8809}{2.9160} = 0.3021, w_2 = \frac{1.3119}{2.9160} = 0.4499$ and $w_3 = \frac{0.7232}{2.9160} = 0.2480$. Compared to the weights by the SDW method (namely, 0.3323, 0.3548 and 0.3129, respectively), both w1 and w3 are lower, whereas w2 is higher; these are due to correlations among the criteria (Table 5.4).

5.7 Gini Coefficient Weighting (GCW) Method

The GCW method, proposed by Li and Chi (2009), quantifies variation in the values of each criterion by the Gini coefficient, which is popular for assessing the distribution and disparity of income and wealth of people in a region or country. It does not require normalization of criteria values and its two steps (presented below) are simple.

Step 1. Compute the Gini coefficient of the *j*th criterion (Gj) by the following equation based on pairwise absolute differences of the discrete values of this criterion:

$$G_j = \frac{\sum_{i=1}^{m}\sum_{k=1}^{m}\left|f_{ij} - f_{kj}\right|}{2m^2 \overline{f}_j} \quad \text{for } j \in (1,2,\ldots,n) \tag{5.11}$$

Here, the term in the double summation is the absolute difference between the original values of the jth criterion for two alternatives. The numerator has m^2 terms (of which m terms are zero), each being an absolute difference; this number of terms justifies m^2 in the denominator.

Finally, $\overline{f}_j = \frac{1}{m}\sum_{i=1}^{m} f_{ij}$ is the arithmetic mean of the original values of the jth criterion. According to Eq. 5.11, the Gini coefficient is half of the relative mean of absolute differences and its value is between 0.0 and 1.0.

NUMERICAL CALCULATIONS (USING ACM FROM TABLE 5.1)

Consider the original values of C2 in Table 5.1, where $m = 5$. In this case, $j = 2$ and f_{i2} values are 2.33, 1.08, 6.45, 8.88 and 7.39 and their mean is $\overline{f}_2 = \frac{1}{5}(2.33+1.08+6.45+8.88+7.39) = 5.226$. Then, using Eq. 5.11,

$$G_2 = \frac{\begin{bmatrix} \left[|2.33-2.33|+|2.33-1.08|+|2.33-6.45|+|2.33-8.88|+|2.33-7.39|\right]+ \\ \left[|1.08-2.33|+|1.08-1.08|+|1.08-6.45|+|1.08-8.88|+|1.08-7.39|\right]+ \\ \left[|6.45-2.33|+|6.45-1.08|+|6.45-6.45|+|6.45-8.88|+|6.45-7.39|\right]+ \\ \left[|8.88-2.33|+|8.88-1.08|+|8.88-6.45|+|8.88-8.88|+|8.88-7.39|\right]+ \\ \left[|7.39-2.33|+|7.39-1.08|+|7.39-6.45|+|7.39-8.88|+|7.39-7.39|\right] \end{bmatrix}}{2\times 5^2 \times 5.226}$$

$$= 0.3163$$

Similar calculations for criteria, C1 and C3, give $G_1 = 0.2836$ and $G_3 = 0.2738$. As expected, these Gini coefficients are less than unity.

Step 2. Now, using G_j values from the previous step, calculate the weight of each criterion by:

$$w_j = \frac{G_j}{\sum_{k=1}^{n} G_k} \quad \text{for } j \in (1,2,\ldots,n) \tag{5.12}$$

This normalization of G_j values is like Eqs. 5.2, 5.4 and 5.10 in the SDW, SVW and CRITIC methods, respectively.

NUMERICAL CALCULATIONS (USING ACM FROM TABLE 5.1)

Calculations in Step 4 gave $G_1 = 0.2836$, $G_2 = 0.3163$ and $G_3 = 0.2738$, whose sum is 0.8737. Hence, the weights of C1, C2 and C3 are, respectively, $w_1 = \frac{0.2836}{0.8737} = 0.3246, w_2 = \frac{0.3163}{0.8737} = 0.3620$ and $w_3 = \frac{0.2738}{0.8737} = 0.3134$. These will be compared later with those weights found by other methods in this chapter.

5.8 Method Based on the Removal Effects of Criteria (MEREC)

MEREC was proposed by Keshavarz-Ghorabaee et al. (2021); it takes a different approach for computing criteria weights using ACM data. As the name indicates, the criteria weights are computed considering the effect of removing one criterion at a time on the overall performance of each alternative, which is quantified through a nonlinear equation. Steps in the MEREC are described below.

Step 1. Normalize ACM data using the following equations, as in Keshavarz-Ghorabaee et al. (2021):

$$F_{ij} = \frac{\underset{k \in \{1,2,\ldots,m\}}{min} f_{kj}}{f_{ij}} \quad \text{for all benefit criteria} \tag{5.13a}$$

$$F_{ij} = \frac{f_{ij}}{\underset{k \in \{1,2,\ldots,m\}}{max} f_{kj}} \quad \text{for all cost criteria} \tag{5.13b}$$

In Eq. 5.13a, the numerator ($\underset{k \in \{1,2,\ldots,m\}}{min} f_{kj}$) is the minimum of the values of the benefit criterion across all alternatives. Eq. 5.13a transforms all benefit criteria to cost criteria (i.e., normalized values of the criterion should be minimized). In Eq. 5.13b, the denominator ($\underset{k \in \{1,2,\ldots,m\}}{max}$) is the maximum of the values of the cost criterion across all m alternatives. Eq. 5.13b does not change the minimization of cost criteria. Hence, all normalized values by Eq. 5.13 should be minimized. Eqs. 5.13a and 5.13b are opposite to those for max normalization (Eqs. 4.1a and 4.1b) and so they can be termed min normalization.

NUMERICAL CALCULATIONS (USING ACM FROM TABLE 5.1)

For instance, for the maximization criterion, C1 ($j = 1$), minimum of its values (namely, 0.185, 0.317, 0.555, 0.731 and 0.948) in the given ACM is 0.185. Hence, the normalized value of C1 for the second alternative is $\frac{0.185}{0.317} = 0.5836$. Normalized values of C1 for all five alternatives are summarized in the second column of Table 5.5.

On the other hand, for the minimization criterion, C2 ($j = 2$), maximum of its values (namely, 2.33, 1.08, 6.45, 8.88 and 7.33) in the given ACM is 8.88. Hence, the normalized value of C2 for the first alternative is $\frac{2{,}33}{8.88} = 0.2624$. Normalized values of C2 for all alternatives are summarized in the third column of Table 5.5, wherein values of C3 are also after min normalization.

TABLE 5.5

Values of ACM in Table 5.1 After Min Normalization.

Alternatives	C1	C2	C3
A1	1.0000	0.2624	0.3833
A2	0.5836	0.1216	0.5839
A3	0.3333	0.7264	1.0000
A4	0.2531	1.0000	0.2049
A5	0.1951	0.8322	0.3366

Step 2. Estimate the overall performance, P_i, of the ith alternative, considering all criteria in ACM, by the following nonlinear equation involving two natural logarithms.

$$P_i = ln\left[1 + \frac{1}{n}\sum_{j=1}^{n}\left|ln\left(F_{ij}\right)\right|\right] \quad \text{for } i \in (1,2,\ldots,m) \tag{5.14}$$

Assuming all values in the original ACM (Table 5.1) are positive, normalized values (F_{ij}) will be > 0.0 and ≤ 1.0 (as in Table 5.5). Hence, $ln\left(F_{ij}\right)$ will be negative. Since the argument of the inner logarithm is an absolute value, the summation in Eq. 5.14 will be more than 0.0 and P_i will be positive. As illustrated by Keshavarz-Ghorabaee et al. (2021), the equation, $\ln\left[1 + \left|\ln(x)\right|\right]$, has large values for small x values. For example, for x = 0.01, 0.05, 0.1, 0.2, 0.4 and 0.8, this equation gives 1.72, 1.39, 1.19, 0.96, 0.65 and 0.20, respectively (i.e., nonlinear for x = 0.0–0.2 and nearly linear for x > 0.2).

NUMERICAL CALCULATIONS (USING ACM FROM TABLE 5.1)

Consider the normalized ACM values in Table 5.5, where $m = 5$ and $n = 3$. Then, for alternative A1 (i.e., $i = 1$), Eq. 5.14 gives:

$$P_1 = ln\left[1 + \frac{1}{3}\left\{\left|\ln(1.0)\right| + \left|\ln(0.2624)\right| + \left|\ln(0.3833)\right|\right\}\right] = 0.5685.$$

Likewise, $P_2 = 0.7233$, $P_3 = 0.3872$, $P_4 = 0.6863$ and $P_5 = 0.6775$.

Step 3. Estimate the overall performance, P_{ij}, of the ith alternative as in Step 1 but excluding the jth criterion in ACM. The modified equation for this calculation is:

$$P_{ij} = ln\left[1 + \frac{1}{n}\sum_{k=1,k\neq j}^{n}\left|ln\left(F_{ik}\right)\right|\right] \quad \text{for } i \in (1,2,\ldots,m) \tag{5.15}$$

Note the exclusion of the jth criterion in the summation. This calculation is repeated by excluding each criterion in ACM, that is, $j \in (1,2,\ldots,n)$, one by one. The result of these calculations will give an (m×n) matrix of values for P_{ij} for $i \in (1,2,\ldots,m)$ and $j \in (1,2,\ldots,n)$.

NUMERICAL CALCULATIONS (USING ACM FROM TABLE 5.1)

Continuing the illustration from the previous step and using normalized ACM values in Table 5.5, calculation of P_{ij} by Eq. 5.15 is illustrated for alternative A2 (i.e., $i = 2$) and removal of one criterion: C1 or $j = 1$.

$$P_{21} = ln\left[1 + \frac{1}{3}\{|(\ln(0.1216)| + |\ln(0.5839)|\}\right] = 0.6321$$

Likewise, P_{ij} for $i \in (1,2,3,4,5)$ and $j \in (1,2,3)$ are calculated and the results are given in Table 5.6.

TABLE 5.6

Overall Performance of Alternatives After Removal of a Criterion.

Alternatives	C1 Removed	C2 Removed	C3 Removed
A1	0.5685	0.2774	0.3688
A2	0.6321	0.3066	0.6322
A3	0.1013	0.3120	0.3872
A4	0.4242	0.6863	0.3771
A5	0.3536	0.6459	0.4737

Step 4. Calculate the overall effect of removing the jth criterion in ACM for all m alternatives together, by:

$$R_j = \sum_{i=1}^{m} |P_{ij} - P_i| \quad \text{for } j \in (1,2,\ldots,n) \tag{5.16}$$

Here, recall that P_{ij} is the overall performance of the ith alternative with j^{th} criterion excluded in ACM, whereas P_i is the overall performance of the ith alternative considering all n criteria. The absolute difference in the summation quantifies the effect of removing j^{th} criterion on the ith alternative.

NUMERICAL CALCULATIONS (USING ACM FROM TABLE 5.1)

Using Eq. 5.16 along with $P_1 = 0.5685$, $P_2 = 0.7233$, $P_3 = 0.3872$, $P_4 = 0.6863$ and $P_5 = 0.6775$ from Step 2 and P_{ij} values in Table 5.6, the calculation of R_1 is as follows:

$$R_1 = |0.5685 - 0.5685| + |0.6321 - 0.7233| + |0.1013 - 0.3872| + |0.4242 - 0.6863| + |0.6775 - 0.3536|$$

$$= 0.9631$$

Similarly, $R_2 = 0.8145$ and $R_3 = 0.8038$.

Step 5. Finally, use R_j values from the previous step to compute the weight of each criterion.

$$w_j = \frac{R_j}{\sum_{k=1}^{n} R_k} \quad \text{for } j \in (1, 2, \ldots, n) \tag{5.17}$$

This step is like the last step in many other methods such as SDW, SVW, CRITIC and GCW.

NUMERICAL CALCULATIONS (USING ACM FROM TABLE 5.1)

Calculations in Step 4 gave $R_1 = 0.9631$, $R_2 = 0.8145$ and $R_3 = 0.8038$, whose sum is 2.5812. So, the weights of C1, C2 and C3 by Eq. 5.17 are, respectively, $w_1 = \frac{0.9631}{2.5812} = 0.3731, w_2 = \frac{0.8145}{2.5812} = 0.3156$ and $w_3 = = \frac{0.,8038}{2.5812} = 0.3114$. These will be compared later with those weights found by other methods in this chapter.

5.9 Discussion

Weights found by the seven weighting methods for the ACM data in Table 5.1 are compared in Table 5.7. Clearly, weights by different methods are different, although some of them are comparable. Among the three criteria, C2 has a higher weight by five out of six methods (excluding the MW method, which gives the same weight), whereas MEREC gave a higher weight for C1. Weight

TABLE 5.7

Weights for the Three Criteria in ACM in Table 5.1, by Seven Methods.

Method	C1	C2	C3
MW	0.3333	0.3333	0.3333
SDW	0.3323	0.3548	0.3129
SVW	0.3305	0.3767	0.2929
EW	0.2973	0.4279	0.2748
CRITIC	0.3021	0.4499	0.2480
GCW	0.3246	0.3620	0.3134
MEREC	0.3731	0.3156	0.3114

for C2 varies from 0.3156 to 0.4499 (i.e., by 35%). On the other hand, criterion C3 has a lower weight, from 0.2480 to 0.3333 (i.e., varies by 23%). The weight of the C2 criterion varies by 29% from 0.2480 to 0.3333. These results indicate potential variation in weights of criteria by different weighting methods; however, they cannot be generalized since variation in weights depends on both ACM data and the weighting methods tested.

Normalization of ACM data is the first step in many objective weighting methods (e.g., SDW, SVW, EW, CRITIC and MEREC); it is not required in a few methods such as MW and GCW. In this chapter, the normalization method in SDW, SVW, EW, CRITIC and MEREC is as in the original/cited reference for each of them. Research shows that normalization, if required in an objective weighting method, affects the computed weights of criteria (Baydaş et al., 2024). For example, Wang et al. (2020) studied the effect of normalization methods on criteria weights by SDW, SVW, EW and CRITIC methods for two MCDM applications. Their results for one ACM involving three criteria are summarized in Table 5.8. These confirm that the calculated values of criteria weights depend on the normalization and weighting methods; the variation can be small or large.

Mukhametzyanov (2021) studied the effect of normalization, SDW, EW and CRITIC methods on criteria weights. He concluded that the max–min method and its generalization are preferable; one reason is that the range of normalized values is always the same, from 0.0 to 1.0. Also, Mukhametzyanov demonstrated that the EW method is highly sensitive and proposed two modifications to this method; one modification is inclusion of correlation coefficients into the EW method (like in the CRITIC method).

Computational effort for finding criteria weights by objective weighting methods (such as those described in this chapter) is not significant in the current era of technology. From the point of manual calculations, some methods such as MW, SDW and SVW are relatively easy compared to the EW, CRITIC, GINI and MEREC methods.

Finally, it is challenging to choose an appropriate or correct weighting method, mainly because the correct weights of criteria in MCDM applications

TABLE 5.8

Effect of Normalization Method on Weights of Criteria by SDW, SVW, EW and CRITIC Methods, for an ACM Having Three Criteria (i.e., MCDM Application).

Weighting Method	Normalization Method	Weight of Criterion 1	Weight of Criterion 2	Weight of Criterion 3
SDW	Max	0.2403	0.3795	0.3802
SDW	Max–Min	0.3321	0.3348	0.3331
SDW	Sum	0.2005	0.3949	0.4046
SDW	Vector	0.2069	0.3922	0.4008
SVW	Max	0.1667	0.4159	0.4175
SVW	Max–Min	0.3308	0.3363	0.3329
SVW	Sum	0.1117	0.4334	0.4549
SVW	Vector	0.1198	0.4305	0.4497
EW	Max	0.2485	0.3808	0.3706
EW	Max–Min	0.3654	0.3243	0.3103
EW	Sum	0.1134	0.4358	0.4508
EW	Vector	0.3441	0.3284	0.3275
CRITIC	Max	0.1964	0.3538	0.4499
CRITIC	Max–Min	0.2535	0.3246	0.4218
CRITIC	Sum	0.1482	0.4769	0.3749
CRITIC	Vector	0.1533	0.4746	0.3721

are unknown. In view of this and the dependency of criteria weights on normalization and weighting methods, objective weighting methods are not exactly "objective" (i.e., not free from bias), since one of them needs to be chosen. In general, several sets of weights should be tried with the chosen ranking methods. So, try several combinations of normalization, weighting and ranking methods for an MCDM application and then review the recommended alternatives for final selection.

5.10 Summary

The main points in this chapter on objective weighting methods are as follows:

1. Weights of criteria are required for ranking by many MCDM methods and there are many weighting methods.
2. This chapter is on objective weighting methods, which generally require ACM data for calculating criteria weights.

3. Objective weighting methods described in this chapter include statistics-based methods such as MW, SDW, SVW and CRITIC and methods based on entropy, the Gini coefficient and the effect of removing a criterion (i.e., EW, GCW and MEREC methods).
4. The first step in almost all objective weighting methods is the normalization of ACM data. Calculated weights of criteria by an objective weighting method depend on the method used for this normalization.
5. In addition, the calculated weights of criteria vary from one objective weighting method to another.
6. Correct values of the criteria weights for MCDM applications are unknown. So, correct normalization and weighting methods are not known.
7. Despite the above three points, one must choose a normalization method and a weighting method for an MCDM application. Hence, objective weighting methods are not exactly "objective" (i.e., not free from bias).
8. In general, one should try several combinations of normalization, weighting and/or ranking methods for an MCDM application and then review the recommended alternatives for different combinations.

References

Baydaş, M., Kavacık, M., & Wang, Z. (2024). Interpreting the determinants of sensitivity in MCDM methods with a new perspective: An application on e-scooter selection with the PROBID method. *Spectrum of Engineering and Management Sciences*, 2(1), 17–35.

Diakoulaki, D., Mavrotas, G., & Papayannakis, L. (1995). Determining objective weights in multiple criteria problems: The critic method. *Computers & Operations Research*, 22(7), 763–770.

Hwang, C.-L., & Yoon, K. (1981). *Multiple attribute decision making: Methods and applications*. Berlin, Germany: Springer.

Keshavarz-Ghorabaee, M., Amiri, M., Zavadskas, E. K., Turskis, Z., & Antucheviciene, J. (2021). Determination of objective weights using a new method based on the removal effects of criteria (MEREC). *Symmetry*, *13*(4), 525.

Li, G., & Chi, G. (2009, 26–28 December). A new determining objective weights method-gini coefficient weight. In *Proceedings of the 2009 First International Conference on Information Science and Engineering*. Nanjing, China.

Mukhametzyanov, I. (2021). Specific character of objective methods for determining weights of criteria in MCDM problems: Entropy, CRITIC and SD. *Decision Making: Applications in Management and Engineering*, 4(2), 76–105.

Nabavi, S. R., Wang, Z., & Rangaiah, G. P. (2023). Sensitivity analysis of multi-criteria decision-making methods for engineering applications. *Industrial & Engineering Chemistry Research*, *62*(17), 6707–6722.

Nabavi, S. R., Wang, Z., & Rangaiah, G. P. (2024). Sensitivity assessment of multi-criteria decision-making methods in chemical engineering optimization applications. In F. Gómez-Castro, & V. Rico-Ramírez (Eds.), *Optimization in chemical engineering: Deterministic, meta-heuristic and data-driven techniques* (pp. 283–304). Berlin, Boston: De Gruyter. https://doi.org/10.1515/9783111383439-008

Rao, R. V., & Patel, B. K. (2010). A subjective and objective integrated multiple attribute decision making method for material selection. *Materials & Design*, *31*(10), 4738–4747.

Wang, Y., Wang, Z., & Wu, Z. (2025). Multi-objective optimal control of nonlinear processes using reinforcement learning with adaptive weighting. *Computers & Chemical Engineering*, 109206.

Wang, Z., Parhi, S. S., Rangaiah, G. P., & Jana, A. K. (2020). Analysis of weighting and selection methods for Pareto-optimal solutions of multiobjective optimization in chemical engineering applications. *Industrial & Engineering Chemistry Research*, *59*(33), 14850–14867.

Wang, Z., Nabavi, S. R., & Rangaiah, G. P. (2023). Selected multi-criteria decision-making methods and their applications to product and system design. In *Optimization Methods for Product and System Design* (pp. 107–138). Singapore: Springer Nature Singapore.

Wang, Z., Nabavi, S. R., & Rangaiah, G. P. (2024). Multi-criteria decision making in chemical and process engineering: Methods, progress and potential. *Processes*, *12*(11), 2532.

Exercises

E5.1 Instead of max–min normalization employed for illustrating the SDW method in Section 5.3, apply (1) sum normalization and (2) max normalization to find the criteria weights by the SDW method. Compare the weights found with those given in Section 5.3.

E5.2 Instead of max–min normalization employed for illustrating the SVW method in Section 5.4, apply (1) sum normalization and (2) vector normalization to find the criteria weights by the SVW method. Compare the weights found with those given in Section 5.4.

E5.3 Instead of sum normalization employed for illustrating the EW method in Section 5.5, apply (1) max–min normalization and (2) max normalization to find the criteria weights by the EW method. Compare the weights found with those given in Section 5.5.

E5.4 Instead of max–min normalization employed for illustrating the CRITIC method in Section 5.6, apply (1) sum and (2) max normalization to find the criteria weights by the CRITIC method. Compare the weights found with those given in Section 5.6.

E5.5 Which of the normalization methods in Chapter 4 can be used in each of the SDW, SVW and CRITIC methods for normalization of ACM data? Analyze and justify.

E5.6 Which of the normalization methods in Chapter 4 can be used in each of the EW and MEREC methods for normalization of ACM data? Analyze and justify.

6

Subjective Weighting Methods

6.1 Overview

In multi-criteria decision-making (MCDM), determining the weight value for each criterion is crucial. For this, there are two main categories of weighting methods, namely, objective and subjective methods. The former methods are described in the previous chapter, whereas the latter methods are detailed in the present chapter. Unlike objective methods, subjective weighting methods do not require the criteria values of alternatives and the number of alternatives. Instead, they require just the inputs on the relative importance of criteria (e.g., in the form of pairwise comparisons of criteria) from decision-makers (Wang et al., 2024). Hence, the computed weights will depend on these inputs provided by decision-makers. Further, the inputs and computed weights will be subjective depending on the evaluation of decision-makers.

The present chapter delves into four popular subjective weighting methods, detailing their principles and equations; these methods are presented in alphabetical sequence. Each method is elucidated with a numerical example, ensuring readers gain a clear understanding of its inner workings. The remaining sections in this chapter are as follows:

Section 6.2	Analytic Hierarchy Process (AHP) Method
Section 6.3	Best–Worst Method (BWM)
Section 6.4	Rank-Ordering Methods (ROMs)
Section 6.5	Stepwise Weight Assessment Ratio Analysis (SWARA) Method
Section 6.6	Discussion
Section 6.7	Summary

The learning outcomes of this chapter are to (1) describe the subjective weighting methods covered in this chapter and (2) choose and apply any of them in decision-making problems.

DOI: 10.1201/9781003530435-6

Unless otherwise stated, the notation used throughout this chapter is as follows:

- i and $j \in \{1, 2, \dots, n\}$ are the indices for criteria in the MCDM problem
- n is the number of criteria in the MCDM problem.
- w_j is the weight assigned to the jth criterion and $\sum_{j=1}^{n} w_j = 1$.

6.2 Analytic Hierarchy Process (AHP) Method

The AHP method, introduced by Saaty (1977), relies on decision-makers making subjective pairwise comparisons, for which Saaty's fundamental 1–9 scale (in Table 6.1) is commonly used. The method starts with the creation of a decision hierarchy, structuring the main goal, n criteria (then sub-criteria, if any) and m alternatives into a model, as shown in Figure 6.1. Decision-makers then use Saaty's 1–9 scale to evaluate the relative importance between pairs of criteria through subjective pairwise comparisons. These evaluations are assembled into an $n \times n$ pairwise comparison matrix. The priority vector (criteria weights) is then derived using the eigenvalue method (Saaty, 1990). Alternatively, a simpler approximation method, based on normalizing the sum of each column in the pairwise comparison matrix, is frequently used in practice (e.g., Karim and Karmaker, 2016; Wang et al., 2020).

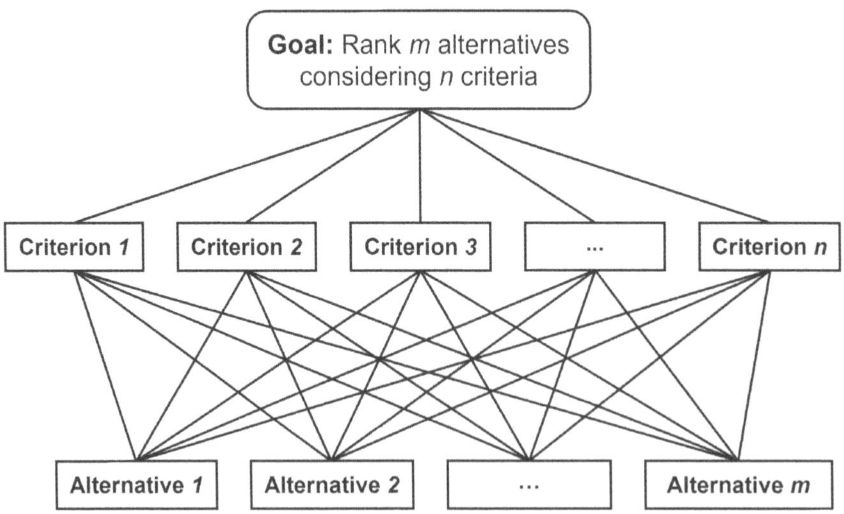

FIGURE 6.1
Decision hierarchy of goal, criteria and alternatives in the AHP method.

TABLE 6.1

Saaty's 1–9 Pairwise Comparison Scale.

Scale	Compare Between Events x and y
1	Equal importance of x against y
3	Moderate importance of x against y
5	Essential or strong importance of x against y
7	Very strong importance of x against y
9	Extreme importance of x against y
2, 4, 6, 8	Intermediate value between adjacent scales
Any decimal within [1, 9], for example, 2.3 and 5.1	Finer scale
Reciprocals, for example, 1/3 and 1/7	If event x has one of the above numbers assigned to it when compared with event y, then y has the reciprocal value when compared with x.

The focus of the AHP method in this section is on deriving criteria weights at the criteria level of the hierarchy in Figure 6.1. A complete walk-through of applying AHP as an MCDM method to rank alternatives is presented in Section 8.4 of Chapter 8. In other words, the AHP concept is employed for both criteria weights and ranking the alternatives in the MCDM field. The main steps of the AHP method for determining criteria weights are outlined below.

Step 1. Construct the $n \times n$ pairwise comparison matrix for n criteria as follows:

$$\mathbf{M}_{\mathrm{A}} = \begin{bmatrix} a_{11} & a_{12} & a_{13} & \cdots & a_{1n} \\ a_{21} & a_{22} & a_{23} & \cdots & a_{2n} \\ a_{31} & a_{32} & a_{33} & \cdots & a_{3n} \\ \vdots & \vdots & \vdots & \ddots & \vdots \\ a_{n1} & a_{n2} & a_{n3} & \cdots & a_{nn} \end{bmatrix} \tag{6.1}$$

- Here, $i, j \in \{1, 2, \ldots, n\}$ and each entry a_{ij} represents the relative importance of criterion i compared to criterion j using Saaty's 1–9 scale (in Table 6.1).
- The leading diagonal entries $(a_{11}, a_{22}, a_{33}, \ldots, a_{nn})$ of this matrix are always equal to unity; that is, the relative importance of a criterion compared to itself is always 1.
- The following equation always holds true (i.e., entries below the leading diagonal are the reciprocals of the corresponding entries above it), as per the last row of Table 6.1.

$$a_{ij} = \frac{1}{a_{ji}} \tag{6.2}$$

- For instance, if we consider that the relative importance of criterion 1 versus criterion 2 is moderate (i.e., $a_{12} = 3$), then the relative importance of criterion 2 to criterion 1 is its reciprocal (i.e., $a_{21} = 1/3$). Thus, $n(n-1)/2$ comparisons are needed to populate half of the matrix and then values of the other half are obtained through simple reciprocals.

NUMERICAL CALCULATIONS

Suppose that the decision-maker provides the following subjective assessments for the three considered criteria (C1, C2 and C3) using Saaty's 1–9 scale:

- C1 is moderately more important than C2 (i.e., $a_{12} = 3$ and $a_{21} = 1/3$), meaning C1 is three times as important as C2, while C2 holds only one-third the importance of C1.
- C1 is strongly more important than C3 (i.e., $a_{13} = 5$ and $a_{31} = 1/5$).
- C2 is rated in-between moderate and strong in importance relative to C3 (i.e., $a_{23} = 4$ and $a_{32} = 1/4$).

Based on these assessments, the 3×3 pairwise comparison matrix is as follows:

$$\mathbf{M_A} = \begin{bmatrix} a_{11} & a_{12} & a_{13} \\ a_{21} & a_{22} & a_{23} \\ a_{31} & a_{32} & a_{33} \end{bmatrix} = \begin{bmatrix} 1 & 3 & 5 \\ 1/3 & 1 & 4 \\ 1/5 & 1/4 & 1 \end{bmatrix}$$

Step 2. Compute the priority vector (i.e., criteria weights in this step), using the eigenvalue method by solving the following equation.

$$\mathbf{M_A}\,\mathbf{w}^* = \lambda_{max}\,\mathbf{w}^* \tag{6.3}$$

Here, λ_{max} is the largest eigenvalue (known as the principal eigenvalue) of the pairwise comparison matrix and $\mathbf{w}^*$ is the corresponding principal eigenvector. The entries of this principal eigenvector are then normalized to sum to 1, yielding the criteria weights, as follows:

$$w_j = \frac{w_j^*}{\sum_{k=1}^{n} w_k^*},\ j = 1,2,\ldots,n \tag{6.4}$$

Here, w_j^* and w_k^* represent the jth and k^{th} entries of the principal eigenvector $\mathbf{w}^*$, respectively, whereas w_j is the weight of the criterion j.

NUMERICAL CALCULATIONS

Solve the following equation using computational tools such as WolframAlpha (https://www.wolframalpha.com/input?i=eigenvalue+calculator) or Python code (Table 6.2), both freely available:

$$\begin{bmatrix} 1 & 3 & 5 \\ 1/3 & 1 & 4 \\ 1/5 & 1/4 & 1 \end{bmatrix} \mathbf{w}^* = \lambda_{\max} \mathbf{w}^*$$

Use of WolframAlpha gives the principal eigenvalue $\lambda_{\max} = 3.086$ and the corresponding principal eigenvector $\mathbf{w}^* = \begin{bmatrix} 6.6943 \\ 2.9876 \\ 1 \end{bmatrix}$. When different computational tools are utilized, the computed principal eigenvector may be scaled by a constant factor; however, this will not affect the final weights due to normalization by Eq. 6.4.

Next, apply sum normalization to the principal eigenvector to derive the weights of the three criteria (C1, C2 and C3), as follows:

$$w_1 = \frac{w_1^*}{\sum_{k=1}^{3} w_k^*} = \frac{6.6943}{6.6943 + 2.9876 + 1} = 0.6267$$

$$w_2 = \frac{w_2^*}{\sum_{k=1}^{3} w_k^*} = \frac{2.9876}{6.6943 + 2.9876 + 1} = 0.2797$$

$$w_3 = \frac{w_3^*}{\sum_{k=1}^{3} w_k^*} = \frac{1}{6.6943 + 2.9876 + 1} = 0.0936$$

TABLE 6.2
Sample Python Code to Calculate Criteria Weights Using the Eigenvalue Method.

```
import numpy as np

# subjective pairwise comparison matrix from Step 1
arr = np.array([
    [1,    3,   5],
    [1/3,  1,   4],
    [1/5,  1/4, 1]
])

# compute eigenvalues and eigenvectors of the matrix using np
eigenvalues, eigenvectors = np.linalg.eig(arr)

# find the index of the principal (maximum) eigenvalue
max_idx = np.argmax(eigenvalues)

# extract the corresponding principal eigenvector
principal_eigenvector = eigenvectors[:, max_idx].real

# sum-normalize the principal eigenvector to get the weights
weights = principal_eigenvector / principal_eigenvector.sum()

# display weights
print("Criteria Weights:")
for i, w in enumerate(weights, 1):
    print(f"w{i} = {w:.4f}")

#-----------output-----------
# Criteria Weights:
# w1 = 0.6267
# w2 = 0.2797
# w3 = 0.0936
#---------------------------
```

As stated earlier, these weights of criteria depend on the values given by decision-makers for pairwise comparison of criteria (i.e., for $\mathbf{M}_A$)

Step 3. Check the consistency of the pairwise comparison matrix of n criteria by calculating its consistency ratio (CR).

$$CI = \frac{\lambda_{max} - n}{n-1},\ CR = \frac{CI}{RI} \tag{6.5}$$

Here, CI is the consistency index and RI (given in Table 6.3) is the random CI value provided by Saaty (1990), who computed it by

TABLE 6.3

Saaty's RI Values for $n \leq 10$. See Podvezko (2009) for $n > 10$.

n	1	2	3	4	5	6	7	8	9	10
RI	0	0	0.58	0.90	1.12	1.24	1.32	1.41	1.45	1.49

averaging the CI values from numerous randomly generated $n \times n$ reciprocal matrices using Saaty's 1–9 scale. If $CR \leq 0.1$, consistency is considered acceptable. If not, decision-makers should review and revise their given values for pairwise comparison of criteria. In addition, for a perfectly consistent pairwise comparison matrix (where each comparison a_{ij} between criterion i and criterion j exactly equals the ratio of their resulting weights w_i / w_j, which also implies that the transitive property holds, i.e., $a_{ij}a_{jk} = a_{ik}$ for all criteria i, j and k), Saaty (1990) proved that the largest eigenvalue of this perfectly consistent pairwise comparison matrix is equal to the number of criteria, that is, $\lambda_{max} = n$.

Note on Consistency Analysis: Because the pairwise comparison matrix is based on human judgment, it is inherently subjective and may occasionally be vague or inconsistent. For instance, if a decision-maker rates criterion 1 as twice as important as criterion 2 and criterion 2 as four times more important than criterion 3, then logically, criterion 1 should be rated eight times more important than criterion 3. This logical relationship is known as the transitive property. However, in real-world scenarios and when the number of criteria is large, decision-makers may unintentionally violate this principle due to cognitive biases or inconsistencies in judgment.

NUMERICAL CALCULATIONS

In this example, with $n = 3$, refer to Table 6.3 and find the corresponding $RI = 0.58$.

$$CI = \frac{\lambda_{max} - n}{n-1} = \frac{3.086 - 3}{3-1} = 0.0429$$

$$CR = \frac{CI}{RI} = \frac{0.0429}{0.58} = 0.074$$

Since the calculated CR satisfies $CR \leq 0.1$, the decision-maker's subjective assessments for the three criteria are considered consistent and acceptable.

Simpler Approximation for Weights: Instead of using the eigenvalue method to determine criteria weights in Step 2, a simpler approximation technique based on sum normalization of each column in the pairwise comparison matrix is often used for quick estimation (Wang et al., 2023). For this, perform column normalization of the comparison matrix, $\mathbf{M}_A$, by:

$$A_{ij} = \frac{a_{ij}}{\sum_{k=1}^{n} a_{kj}} \tag{6.6}$$

Then, find weights by row normalization of the matrix of A_{ij} values by:

$$w_i = \frac{\sum_{j=1}^{n} A_{ij}}{n} \tag{6.7}$$

In the above equations, as in the previous steps, the ranges of $i, j \in \{1, 2, \ldots, n\}$ apply.

NUMERICAL CALCULATIONS

For instance, for $i = 2$ and $j = 1$ in the pairwise comparison matrix from Step 1:

$$A_{21} = \frac{a_{21}}{\sum_{k=1}^{3} a_{k1}} = \frac{1/3}{1 + 1/3 + 1/5} = 0.2174$$

Following this approach, the complete column-normalized pairwise comparison matrix is:

$$A_{ij} = \begin{bmatrix} 0.6522 & 0.7059 & 0.5 \\ 0.2174 & 0.2353 & 0.4 \\ 0.1304 & 0.0588 & 0.1 \end{bmatrix}$$

Next, the approximate weight of each criterion is calculated by averaging each row:

$$w_1 = \frac{\sum_{j=1}^{3} A_{1j}}{3} = \frac{0.6522 + 0.7059 + 0.5}{3} = 0.6194$$

$$w_2 = \frac{\sum_{j=1}^{3} A_{2j}}{3} = \frac{0.2174 + 0.2353 + 0.4}{3} = 0.2842$$

$$w_3 = \frac{\sum_{j=1}^{3} A_{3j}}{3} = \frac{0.1304 + 0.0588 + 0.1}{3} = 0.0964$$

Note that these weights automatically satisfy the requirement: $\sum_{j=1}^{n} w_j = 1$.

As can be seen, the above approximate weights are relatively close to those derived using the eigenvalue method. However, this may not always be the case. The difference between the two sets of weights can become significant when the subjective pairwise comparison matrix is highly inconsistent or when the number of criteria increases, which can amplify the effects of judgmental inconsistencies.

6.3 Best–Worst Method (BWM)

Rezaei (2015) introduced the BWM, a nonlinear min–max optimization model that determines criteria weights based on two vectors of pairwise comparisons. This model can yield multiple sets of weights because its optimization is based on ratios between weights. To overcome this issue, Rezaei (2016) proposed a linear version of the method, which guarantees a unique set of weights as the solution. Like the AHP method, BWM derives criteria weights solely from the subjective comparisons between criteria, which are provided by decision-makers. The steps in the linear BWM are described below.

Step 1. Identify the best (i.e., the most preferred or important) criterion, B and the worst (i.e., the least preferred or important) criterion, W, among all the n criteria.

NUMERICAL CALCULATIONS

Suppose there are three criteria (C1, C2 and C3) under consideration and the decision-maker provides the following inputs:

- C1 is identified as the best (the most preferred) criterion, so $B = 1$.
- C3 is identified as the worst (the least preferred) criterion, so $W = 3$.

In effect, C1, C2 and C3 are in decreasing order of preference. If this is not the case, the criteria can be rearranged in decreasing order of preference.

Step 2. Evaluate the preference of the best criterion over each of the remaining criteria, using an integer scale from 1 to 9, where 1 denotes equal importance and 9 represents extreme importance (i.e., nine times more important). This yields the Best-to-Others (BO) comparison vector:

$$A_B = \{a_{B1}, a_{B2}, \ldots, a_{Bn}\} \tag{6.8}$$

Here, a_{Bj} (integer and $1 \leq a_{Bj} \leq 9$) represents the degree to which the best criterion B is preferred over criterion j. Note that $a_{BB} = 1$ (i.e., equal importance when comparing to itself) and the values of a_{Bj} need not be distinct across all j.

NUMERICAL CALCULATIONS

The decision-maker provides the following BO comparison vector:

$$A_B = \{a_{B1}, a_{B2}, a_{B3}\} = \{1, 3, 5\}$$

That is, the best criterion, C1, is considered three times as important as C2 and five times as important as C3 (the worst criterion).

Step 3. Provide the preference of each of the other criteria over the worst criterion using an integer scale from 1 to 9 (i.e., equal importance to extreme importance). This yields the Others-to-Worst (OW) comparison vector:

$$A_W = \{a_{1W}, a_{2W}, \ldots, a_{nW}\} \tag{6.9}$$

Here, a_{jW} (integer and $1 \leq a_{jW} \leq 9$) represents the preference of the criterion j over the worst criterion W. Further, $a_{WW} = 1$ and the values of a_{jW} need not be different for each criterion.

NUMERICAL CALCULATIONS

The decision-maker provides the following OW comparison vector:

$$A_W = \{a_{1W}, a_{2W}, a_{3W}\} = \{5, 2, 1\}$$

As before, the best criterion, C1, is considered five times as important as C3 (the worst criterion). In addition, C2 is considered twice as important as the worst criterion, C3.

From Steps 2 and 3, it can be observed that the total number of comparisons required by BWM is $2n-3$ (i.e., one comparison between the best and worst criteria, $n-2$ comparisons of the best criterion to the other criteria and another $n-2$ comparisons of the other criteria to the worst criterion).

Step 4. The weights of the criteria are determined by solving the following optimization problem.

$$\text{Minimize} \quad \xi \tag{6.10}$$

$$\text{With respect to} \quad w_1, w_2, \ldots, w_n \text{ and } \xi \tag{6.11}$$

$$\text{Subject to} \quad \sum_{j=1}^{n} w_j = 1, \quad w_j \geq 0, \quad \xi \geq 0 \tag{6.12}$$

$$\left| w_B - a_{Bj} w_j \right| \leq \xi, \quad j \in \{1, 2, \ldots, n\} \tag{6.13}$$

$$\left| w_j - a_{jW} w_W \right| \leq \xi, \quad j \in \{1, 2, \ldots, n\} \tag{6.14}$$

In this formulation, ξ represents the maximum allowable deviation from consistency in the pairwise comparisons. A smaller value of ξ indicates a higher level of consistency in the decision-maker's judgments. This optimization problem, after transformation into a linear problem (detailed in the numerical calculations below), can be efficiently solved using computational tools such as Solver in Microsoft Excel, Python and MATLAB® that support constrained linear optimization.

NUMERICAL CALCULATIONS

Taking the comparison vectors (A_B and A_W) from Steps 2 and 3, we can formulate the optimization problem as follows:

$$\text{Minimize} \quad \xi$$

$$\text{With respect to} \quad \text{With respect to } w_1, w_2, w_3 \text{ and } \xi$$

$$\text{Subject to} \quad \sum_{j=1}^{3} w_j = 1, \quad w_j \geq 0, \quad \xi \geq 0$$

$$|w_1 - 1.0\,w_1| \le \xi$$

$$|w_1 - 3.0\,w_2| \le \xi$$

$$|w_1 - 5.0\,w_3| \le \xi$$

$$|w_1 - 5.0\,w_3| \le \xi$$

$$|w_2 - 2.0\,w_3| \le \xi$$

$$|w_3 - 1.0\,w_3| \le \xi$$

Strictly speaking, the above formulation is not yet an LP problem, which requires all constraints and the objective function to be linear expressions of decision variables. A linear expression is one in which each variable appears with a constant coefficient and no variable is multiplied by another variable, raised to a power, placed in a denominator, or wrapped inside a nonlinear function (such as an absolute value, square root, exponential, or trigonometric term). Geometrically, this means that all constraints in an LP must define straight lines or flat hyperplanes in the space of decision variables and the objective function's level sets are parallel hyperplanes that "slide" across this feasible region, so the optimal solution will always lie at one of the region's corner points (vertices) of the resulting convex polyhedron.

However, in the above problem, constraints involving absolute values, such as $|w_1 - 3.0\,w_2| \le \xi$, are only piecewise linear and V-shaped, unlike the straight lines or flat planes described by typical linear constraints. Nonetheless, such constraints can be easily transformed into LP form. Each absolute value inequality can be rewritten as two linear inequalities. For example:

$$|w_1 - 3.0\,w_2| \le \xi \Leftrightarrow \begin{cases} w_1 - 3.0\,w_2 \le \xi \\ -(w_1 - 3.0\,w_2) \le \xi \end{cases}$$

After this transformation, the problem fits the LP structure and can be solved using the LP techniques. For example, it can now be solved using the Simplex method in the Solver tool in Microsoft Excel, as illustrated in Figure 6.2. Alternatively, it can also be solved efficiently using Python libraries such as cvxpy, as demonstrated in Table 6.4.

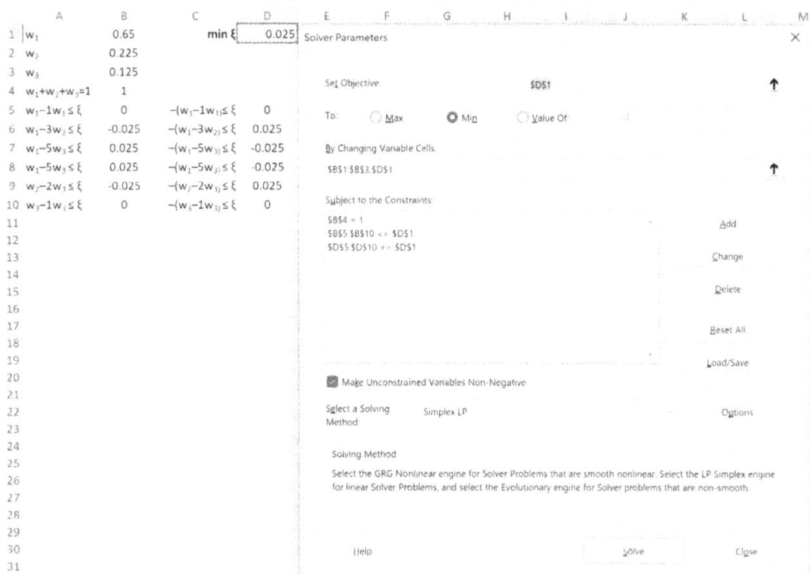

FIGURE 6.2
Screenshot showing the BWM optimization problem in Microsoft Excel and its solution using the Simplex method in the Solver tool.

TABLE 6.4

Sample Python Code Using the cvxpy Library (Version 1.6.6) to Solve the BWM Optimization Problem.

```
# uncomment the next line of code to install cvxpy
# !pip install cvxpy
import cvxpy as cp
import numpy as np

# w.r.t the variables
w1 = cp.Variable()
w2 = cp.Variable()
w3 = cp.Variable()
xi = cp.Variable()

# minimize obj func ξ (xi)
objective = cp.Minimize(xi)

# list the constraints
constraints = [
    w1 + w2 + w3 == 1,
    w1 >= 0,
    w2 >= 0,
    w3 >= 0,
    xi >= 0,
```

(Continued)

(Keršuliene et al., 2010). The final weight for each criterion is obtained by normalizing the Q_j values across all criteria. The SWARA method consists of the following steps.

Step 1. Arrange the criteria in descending order of importance as determined by the decision-maker. The most important criterion is listed first and the least important criterion is listed last.

NUMERICAL CALCULATIONS

Suppose there are six evaluation criteria, denoted as C1, C2, C3, C4, C5 and C6. Each criterion code Cj (where j∈{1,2,3,4,5,6}) can represent a meaningful criterion relevant to a specific decision-making problem; for example, when assessing renewable energy sources, the six criteria may include installation cost, operational cost, efficiency, technical maturity, greenhouse gas emissions and social acceptance.

The decision-maker provides the following order of importance among the criteria:

$$C1 > C2 > C3 = C4 > C5 > C6$$

In effect, C1, C2, C3, C4, C5 and C6 are in decreasing order of preference. If this is not the case, the criteria can be rearranged in decreasing order of preference.

Step 2. Assign a comparative importance value S_j (where $0 \leq S_j \leq 1$) for each criterion j, starting from the second one in the ranked list of criteria. Here, S_j indicates how much more important criterion $j-1$ is compared to criterion j. A value of 0 means they are equally important, while larger values signify a greater importance of criterion $j-1$ over criterion j.

NUMERICAL CALCULATIONS

The decision-maker provides the following S_j value, starting from the second criterion:

$$\{S_2, S_3, S_4, S_5, S_6\} = \{0.1, 0.3, 0.0, 0.4, 0.7\}$$

Note that S_4 is assigned a value of 0.0, since C3 and C4 are considered equally important.

As seen, SWARA requires $n-1$ subjective assessment values from the decision-maker (besides arranging the criteria in the order of decreasing importance).

Step 3. Using the provided S_j values, calculate K_j and Q_j for each criterion ($j = \{1, 2, \ldots, n\}$) using the following equations:

$$K_j = \begin{cases} 1 & \text{if } j = 1 \\ S_j + 1 & \text{if } j > 1 \end{cases} \tag{6.18}$$

$$Q_j = \begin{cases} 1 & \text{if } j = 1 \\ \dfrac{Q_{j-1}}{K_j} & \text{if } j > 1 \end{cases} \tag{6.19}$$

Here, $K_j \geq 1.0$ and $Q_j \leq 1.0$. K_j is the adjusted coefficient of importance for criterion j; it signifies the importance (instead of how much more important) of the previous criterion compared to the current criterion, *j*. Clearly, the more important the previous criterion is compared to the current one (i.e., the higher the S_j), the larger the K_j is. On the other hand, Q_j is the relative weight of the criterion j in relation to the most important criterion. Each Q_j gets progressively smaller (since $K_j \geq 1.0$), meaning lower-ranked criteria get smaller weights.

NUMERICAL CALCULATIONS

For instance, for $j = 2$:

$$K_2 = S_2 + 1 = 0.1 + 1 = 1.1$$

Since $Q_1 = 1$:

$$Q_2 = \frac{Q_1}{K_2} = \frac{1}{1.1} = 0.9091$$

Following this approach, by substituting the given S_j values (in Step 2) into the equations in Step 3, we obtain:

$$\{K_1, K_2, K_3, K_4, K_5, K_6\} = \{1, 1.1, 1.3, 1.0, 1.4, 1.7\}$$

$$\{Q_1, Q_2, Q_3, Q_4, Q_5, Q_6\} = \{1.0, 0.9091, 0.6993, 0.6993, 0.4995, 0.2938\}$$

5. Compared to the AHP method, BWM requires fewer comparisons by focusing only on evaluations relative to the best and worst criteria. It uses an optimization model to compute weights and consistency.
6. ROMs rank the criteria and apply predefined functions to compute criteria weights, offering a simple and intuitive approach without requiring pairwise comparisons or optimization.
7. SWARA ranks criteria in descending order and uses relative importance scores to compute adjustment coefficients, followed by normalization to get criteria weights.
8. The number and type of subjective inputs required vary across these methods, as do their internal mechanisms for consistency or simplification.
9. Choosing a suitable subjective weighting method depends on the complexity of the decision problem, the number of criteria and the level of judgmental effort available from decision-makers.

References

Karim, R., & Karmaker, C. L. (2016). Machine selection by AHP and TOPSIS methods. *American Journal of Industrial Engineering*, *4*(1), 7–13.

Keršuliene, V., Zavadskas, E. K., & Turskis, Z. (2010). Selection of rational dispute resolution method by applying new Step-Wise Weight Assessment Ratio Analysis (SWARA). *Journal of Business Economics and Management*, *11*(2), 243–258.

Pamučar, D., Stević, Ž., & Sremac, S. (2018). A new model for determining weight coefficients of criteria in MCDM models: Full Consistency Method (FUCOM). *Symmetry*, *10*(9), 393.

Podvezko, V. (2009). Application of AHP technique. *Journal of Business Economics and Management*, (2), 181–189.

Rao, R. V. (2024). BHARAT: A simple and effective multi-criteria decision-making that does not need fuzzy logic, part-1: Multi-attribute decision-making applications in the industrial environment. *International Journal of Industrial Engineering Computations*, *15*, 13–40.

Rezaei, J. (2015). Best-worst multi-criteria decision-making method. *Omega*, *53*, 49–57.

Rezaei, J. (2016). Best-worst multi-criteria decision-making method: Some properties and a linear model. *Omega*, *64*, 126–130.

Roszkowska, E. (2013). Rank ordering criteria weighting methods – a comparative overview. *Optimum: Studia Ekonomiczne*, *5*(65), 14–33.

Saaty, T. L. (1977). A scaling method for priorities in hierarchical structures. *Journal of Mathematical Psychology*, *15*(3), 234–281.

Saaty, T. L. (1990). How to make a decision: The analytic hierarchy process. *European Journal of Operational Research*, *48*(1), 9–26.

Stillwell, W. G., Seaver, D. A., & Edwards, W. (1981). A comparison of weight approximation techniques in multiattribute utility decision making. *Organizational Behavior and Human Performance*, *28*(1), 62–77.

Wang, Y., Wang, Z., & Wu, Z. (2025). Multi-objective optimal control of nonlinear processes using reinforcement learning with adaptive weighting. *Computers & Chemical Engineering*, 109206.

Wang, Z., Nabavi, S. R., & Rangaiah, G. P. (2023). Selected multi-criteria decision-making methods and their applications to product and system design. In *Optimization methods for product and system design* (pp. 107–138). Singapore: Springer Nature Singapore. https://doi.org/10.1007/978-981-99-1521-7_7

Wang, Z., Nabavi, S. R., & Rangaiah, G. P. (2024). Multi-criteria decision making in chemical and process engineering: Methods, progress and potential. *Processes, 12*(11), 2532.

Wang, Z., Parhi, S. S., Rangaiah, G. P., & Jana, A. K. (2020). Analysis of weighting and selection methods for Pareto-optimal solutions of multiobjective optimization in chemical engineering applications. *Industrial & Engineering Chemistry Research, 59*(33), 14850–14867.

Exercises

E6.1 Consider a decision-making problem involving five criteria: C1, C2, C3, C4 and C5. You are the decision-maker responsible for determining their weights using the AHP method.

a) Provide your subjective pairwise comparisons and calculate the weights of the five criteria using the eigenvalue method. Also, compute the CR.

b) Based on the CR obtained in part (a), evaluate how consistent your subjective assessments are and whether it is acceptable.

c) Calculate the weights of the five criteria using the approximation method. Compare these weights with those obtained using the eigenvalue method. Discuss the differences, if any.

E6.2 As in Exercise E6.1, consider the same decision-making problem involving five criteria: C1, C2, C3, C4 and C5. Apply the BWM to determine the weights of the criteria based on your subjective assessment values. Compare the resulting weights by BWM with those obtained using the AHP method in E6.1 and discuss the differences observed (in terms of both steps and results).

E6.3 Consider a decision-making problem involving eight criteria. Apply all three ROMs covered in this chapter to determine the weights of the criteria based on your subjective assessments. Clearly outline the steps followed. State your assumptions, if any. Compare the criteria weights found by the three ROMs.

E6.4 Consider a decision-making problem involving eight criteria. Apply the SWARA method to determine the weights of the criteria based on your subjective assessments. Clearly outline the steps followed. Compare the criteria weights found by the SWARA method with those by ROMs in the previous exercise.

E6.5 State whether each of the following statements is correct (right) or incorrect (wrong).

1. Both the AHP method and BWM require ranking of criteria from the most important to the least important.
2. Both the AHP method and BWM require pairwise comparison among all criteria by the decision-maker.
3. In ROMs, criteria weights are computed directly from the assigned ranks using specific transformation functions.
4. ROMs require both the criteria ranks and pairwise comparisons from the decision-maker.
5. SWARA method transforms comparative importance scores into adjustment coefficients, which are then processed and normalized to compute criteria weights.
6. The SWARA method requires ranking of criteria from the most important to the least important and no other input from the decision-makers.

E6.6 Which one of the subjective weighting methods covered in this chapter would you choose and why? State your assumptions of the decision-making problem and justify your choice.

7

Multi-Criteria Decision-Making – Reference-Type Methods

7.1 Overview

Multi-criteria decision-making (MCDM) serves as an effective approach for recommending one or, in some cases, several alternatives from a set of available alternatives, based on values of multiple criteria. As the diverse landscape of MCDM unfolds in this book, this chapter focuses on the reference-type MCDM methods. These methods utilize one or more reference points derived from the ACM of the given MCDM application as a pivotal step. Subsequently, the performance scores of each alternative are calculated based on these reference points. Finally, the alternatives are ranked based on their calculated performance scores.

This chapter delves into nine reference-type MCDM methods, detailing their principles and algorithms. These methods are selected based on either their widespread popularity for various applications or their recent development, allowing the book to remain up to date with the latest advancements in MCDM algorithms. Each method is elucidated through a simple numerical example, ensuring that readers gain a clear understanding of its inner workings.

The following sections in this chapter cover nine reference-type MCDM methods, in chronological order. Readers need not study them sequentially and can read any of them (independent of others).

Section 7.2	Technique for Order of Preference by Similarity to Ideal Solution (TOPSIS)
Section 7.3	Gray/Grey Relational Analysis (GRA)
Section 7.4	VlseKriterijumska Optimizacija I Kompromisno Resenje (VIKOR)
Section 7.5	Evaluation Based on Distance from Average Solution (EDAS)
Section 7.6	Multi-attributive Border Approximation Area Comparison (MABAC)
Section 7.7	Combinative Distance-Based Assessment (CODAS)
Section 7.8	Proximity Indexed Value (PIV)

DOI: 10.1201/9781003530435-7

Similar calculations are performed for all other values in the ACM across the five alternatives and three criteria. The complete results, forming the normalized ACM, are shown in Table 7.2.

TABLE 7.2

Normalized ACM for the TOPSIS Walk-Through.

Alternatives	C1	C2	C3
A1	0.1351	0.1729	0.3962
A2	0.2315	0.0801	0.26
A3	0.4052	0.4785	0.1518
A4	0.5337	0.6588	0.7408
A5	0.6922	0.5483	0.4511

Step 2. Construct the weighted normalized ACM by multiplying the values of each criterion by its assigned weight, w_j.

$$v_{ij} = F_{ij} \times w_j \tag{7.2}$$

NUMERICAL CALCULATIONS

As aforementioned, the assigned weights for the three criteria are: $w_1 = 0.25$, $w_2 = 0.4$ and $w_3 = 0.35$. Then, for $i = 2$ and $j = 1$:

$$v_{21} = F_{21} \times w_1 = 0.2315 \times 0.25 = 0.0579$$

Likewise, calculations are carried out for all other values in the normalized ACM, using their respective assigned weights. The results, which constitute the weighted normalized ACM, are displayed in Table 7.3.

TABLE 7.3

Weighted Normalized ACM for the TOPSIS Walk-Through.

Alternatives	C1	C2	C3
A1	0.0338	0.0691	0.1387
A2	0.0579	0.0321	0.0910
A3	0.1013	0.1914	0.0531
A4	0.1334	0.2635	0.2593
A5	0.1730	0.2193	0.1579

Step 3. From the weighted normalized ACM, determine the PIS, A^+ and NIS, A^-, by the following definitions.

$$\begin{aligned} A^+ &= \left\{ \left(\max_i \left(v_{ij} \right) | j \in J_{max} \right), \left(\min_i \left(v_{ij} \right) | j \in J_{min} \right) \right\} \\ &= \left\{ v_1^+, v_2^+, \ldots, v_j^+, \ldots, v_n^+ \right\} \end{aligned} \tag{7.3}$$

$$\begin{aligned} A^- &= \left\{ \left(\min_i \left(v_{ij} \right) | j \in J_{max} \right), \left(\max_i \left(v_{ij} \right) | j \in J_{min} \right) \right\} \\ &= \left\{ v_1^-, v_2^-, \ldots, v_j^-, \ldots, v_n^- \right\} \end{aligned} \tag{7.4}$$

In the above equations, J_{max} is the set of maximization criteria and J_{min} is the set of minimization criteria, from $\{1, 2, \ldots, n\}$.

NUMERICAL CALCULATIONS

Given that criteria C1 and C3 are to be maximized, we have $J_{max} = \{C1, C3\}$ and, as criterion C2 requires minimization, we have $J_{min} = \{C2\}$.

For instance, for $j = 1$:

$$v_1^+ = \max_i \left(v_{i1} \right) = 0.1730$$

$$v_1^- = \min_i \left(v_{i1} \right) = 0.0338$$

For $j = 2$:

$$v_2^+ = \min_i \left(v_{i2} \right) = 0.0321$$

$$v_2^- = \max_i \left(v_{i2} \right) = 0.2635$$

The PIS is found to be: $A^+ = \left\{ v_1^+, v_2^+, v_3^+ \right\} = \{0.1730, 0.0321, 0.2593\}$. The NIS is found to be: $A^- = \left\{ v_1^-, v_2^-, v_3^- \right\} = \{0.0338, 0.2635, 0.0531\}$. Both these can be easily verified by inspecting values in Table 7.3.

Step 4. Compute the Euclidean distances of each alternative to both PIS and NIS, respectively, by:

$$Distance\, to\, PIS, S_{i+} = \sqrt{\sum_{j=1}^{n} \left(v_{ij} - v_j^+ \right)^2} \tag{7.5}$$

$$Distance\,to\,NIS, S_{i-} = \sqrt{\sum_{j=1}^{n}\left(v_{ij} - v_j^-\right)^2} \tag{7.6}$$

NUMERICAL CALCULATIONS

For instance, for $i = 1$:

$$S_{1+} = \sqrt{\sum_{j=1}^{3}\left(v_{1j} - v_j^+\right)^2}$$

$$= \sqrt{(0.0338 - 0.1730)^2 + (0.0691 - 0.0321)^2 + (0.1387 - 0.2593)^2}$$
$$= 0.1880$$

$$S_{1-} = \sqrt{\sum_{j=1}^{3}\left(v_{1j} - v_j^-\right)^2}$$

$$= \sqrt{(0.0338 - 0.0338)^2 + (0.0691 - 0.2635)^2 + (0.1387 - 0.0531)^2}$$
$$= 0.2124$$

After calculating the Euclidean distances of each alternative to both PIS and NIS, the results are summarized in the first three columns of Table 7.4.

TABLE 7.4

Euclidean Distances of Each Alternative to Both PIS and NIS, as well as Performance Scores (P_i), for the TOPSIS Walk-Through.

Alternatives	S_{i+}	S_{i-}	P_i
A1	0.1880	0.2124	0.5305
A2	0.2039	0.2358	0.5362
A3	0.2703	0.0988	0.2677
A4	0.2348	0.2290	0.4937
A5	0.2130	0.1798	0.4578

Step 5. Compute the performance score (P_i) of each alternative as follows:

$$P_i = \frac{S_{i-}}{S_{i-} + S_{i+}} \tag{7.7}$$

Finally, the alternative with the largest P_i is top-ranked and recommended to the decision-maker.

NUMERICAL CALCULATIONS

For instance, for $i=1$:

$$P_1 = \frac{S_{1-}}{S_{1-}+S_{1+}} = \frac{0.2124}{0.2124+0.1880} = 0.5305$$

The calculated performance scores (presented in the last column of Table 7.4) for all alternatives indicate that A2 > A1 > A4 > A5 > A3, with A2 having the largest score of 0.5362, making it the top-ranked alternative by TOPSIS.

The advantages of TOPSIS are as follows: (1) As one of the earliest MCDM methods, it is widely adopted in many academic and practical applications, including supply chain management, design, engineering, business, health and safety (Behzadian et al., 2012). (2) Using both PIS and NIS, it is intuitive for decision-makers to accept that the top-ranked solution should be closer to the positive ideal while farther away from the negative ideal. (3) Steps in the method are relatively straightforward to understand and implement.

The limitations of TOPSIS are as follows: (1) It uses the square-based Euclidean distance, which may overly penalize large deviations in one criterion. (2) It is more sensitive to outliers; that is, one single extreme alternative can change the PIS and NIS. (3) The addition or removal of alternatives can possibly change the PIS and NIS, leading to the rank reversal phenomenon (i.e., relative rankings of existing alternatives change due to the updated reference points).

7.3 Gray/Grey Relational Analysis (GRA)

The GRA method is founded on the gray system theory proposed by Deng (1982). This theory serves as a powerful analytical tool, particularly in situations where information is incomplete or uncertain (Scarlat, 2015). The core principle of GRA is to evaluate alternatives by measuring their proximity to an ideal reference solution through gray relational coefficients (GRCs). The alternative that is closer to the reference solution receives a higher ranking (Wang et al., 2023).

There are multiple variants of GRA in the MCDM literature. The specific variant referenced here, as outlined in Song and Jamalipour (2005) and Martinez-Morales et al. (2010), is notable for its autonomy from the predefined weights

Step 3. Construct the GRC matrix.

$$GRC_{ij} = \frac{\Delta_{min} + \Delta_{max}}{\Delta_{ij} + \Delta_{max}} \tag{7.12}$$

Where $\Delta_{max} = \max_{i,j}(\Delta_{ij})$ and $\Delta_{\min} = \min_{i,j}(\Delta_{ij})$, which are the global maximum and minimum from the entire difference matrix (Δ), respectively. Note that, by using the max–min normalization method, the above GRC values in any application will be between 0.5 and 1.0 (e.g., see Table 7.7).

NUMERICAL CALCULATIONS

$$\Delta_{max} = \max_{i,j}(\Delta_{ij}) = 1$$

$$\Delta_{min} = \min_{i,j}(\Delta_{ij}) = 0$$

For instance, for $i = 2$ and $j = 1$:

$$GRC_{21} = \frac{\Delta_{min} + \Delta_{max}}{\Delta_{21} + \Delta_{max}} = \frac{0+1}{0.8270+1} = 0.5473$$

The complete GRC matrix is shown in Table 7.7.

TABLE 7.7

GRC Matrix for GRA Walk-Through.

Alternatives	C1	C2	C3
A1	0.5000	0.8619	0.6308
A2	0.5473	1.0000	0.5506
A3	0.6600	0.5923	0.5000
A4	0.7786	0.5000	1.0000
A5	1.0000	0.5528	0.6703

Step 4. Compute the performance score (P_i) of each alternative and the alternative with the largest P_i is top-ranked and recommended to the decision-maker.

$$P_i = \frac{1}{n}\sum_{j=1}^{n} GRC_{ij} \tag{7.13}$$

NUMERICAL CALCULATIONS

For instance, for $i = 2$:

$$P_2 = \frac{1}{3}\sum_{j=1}^{3} GRC_{2j} = \frac{1}{3}(0.5473 + 1.0000 + 0.5506) = 0.6993$$

Similarly, the performance scores for all alternatives are calculated as $P_1 = 0.6642$, $P_2 = 0.6993$, $P_3 = 0.5841$, $P_4 = 0.7595$ and $P_5 = 0.7410$. Accordingly, the ranking is A4 > A5 > A2 > A1 > A3, with A4 being the top-ranked alternative by GRA.

In another widely used GRA variant (Zhang et al., 2011; Kundakçı, 2016), Equation 7.32 in Step 3 is modified as follows:

$$GRC_{ij} = \frac{\Delta_{min} + \zeta\Delta_{max}}{\Delta_{ij} + \zeta\Delta_{max}} \tag{7.14}$$

Here, $\zeta \in [0,1]$ is the distinguishing coefficient, typically set to 0.5 in the literature, providing moderate distinguishing effects and good stability (Özçelik and Öztürk, 2014). Additionally, if the criteria are assigned different weights, Equation 7.33 in Step 4 can be adjusted accordingly, as follows:

$$P_i = \sum_{j=1}^{n} w_j GRC_{ij} \tag{7.15}$$

The advantages of GRA are as follows: (1) The steps in GRA are conceptually simple and intuitive, where a bigger GRC value means an alternative is more like the positive-ideal reference. (2) Some GRA variants, such as the one outlined in Martinez-Morales et al. (2010), can function without requiring explicit criteria weights from decision-makers, reducing the subjectivity. (3) GRA originates from gray system theory (Deng, 1982), which was designed to handle limited, incomplete, or uncertain data. In principle, GRA can be readily adapted to missing or imprecise values. (4) By introducing the distinguishing coefficient $\zeta \in [0,1]$, sensitivity of GRA to differences (Δ's) can be controlled.

The limitations of GRA are as follows: (1) Similar to other reference-type methods, GRA is also susceptible to rank reversal, where the ranking of alternatives may change when an alternative is added or removed. (2) GRA uses only the PIS as the reference point, whereas some decision-making problems might benefit from considering the distances to both the best and the worst (i.e., PIS and NIS). (3) Although the distinguishing coefficient ζ provides flexibility in a GRA variant, deciding its value (e.g., 0.3, 0.5 and 0.7) is arbitrary and the final ranking is sensitive to that choice. There is no fully objective way to set ζ value and so it becomes another subjective input from the decision-makers.

7.4 VlseKriterijumska Optimizacija I Kompromisno Resenje (VIKOR)

The VIKOR (Serbian phrase standing for multi-criteria optimization and compromise solution) method was introduced by Opricovic (1998). It ranks alternatives based on their closeness to the PIS while balancing group benefit (majority rule) and individual regret. The method calculates two key measures: the group utility measure (S), which represents the overall closeness of an alternative to the PIS by aggregating its weighted deviations across all criteria and the individual regret measure (R), which accounts for the worst-performing criterion for each alternative by considering its maximum deviation from the ideal value. An alternative is ranked higher if it is closer to the PIS (i.e., lower S measure) and has a lower R measure. These two measures are then combined into the VIKOR index (Q), which determines the final ranking of the alternatives. The sequential steps in the VIKOR method are as follows:

Step 1. For each criterion, determine its best (f_j^+) and worst (f_j^-) values in the original ACM with m rows (i.e., alternatives) and n columns (i.e., criteria).

$$\text{If the } j^{th} \text{ criterion is for maximization}: f_j^+ = \max_i f_{ij} \text{ and } f_j^- = \min_i f_{ij} \quad (7.16a)$$

$$\text{If the } j^{th} \text{ criterion is for minimization}: f_j^+ = \min_i f_{ij} \text{ and } f_j^- = \min_i f_{ij} \quad (7.16b)$$

NUMERICAL CALCULATIONS (USING ACM FROM TABLE 7.1)

For instance, for $j = 1$, since it is a maximization criterion:

$$f_1^+ = \max_i f_{i1} = 0.948$$

$$f_1^- = \min_i f_{i1} = 0.185$$

Likewise, the best and worst values for the remaining criteria are determined as follows: $f_2^+ = 1.08$, $f_2^- = 8.88$, $f_3^+ = 849$ and $f_3^- = 174$.

Step 2. Normalize the ACM using the following equation, where F_{ij} denotes the normalized deviation of the ith alternative to the PIS at the jth criterion.

$$F_{ij} = \frac{f_j^+ - f_{ij}}{f_j^+ - f_j^-} \tag{7.17}$$

This equation is essentially a slight variation of the conventional max–min normalization. By substituting the expressions for f_j^+ and f_j^-, it becomes evident that the optimization direction for the maximization criteria is reversed. This transformation changes the maximizing f_{ij} to minimizing F_{ij}, due to the negative sign introduced before f_{ij} in the numerator.

NUMERICAL CALCULATIONS

For instance, for $i = 1$ and $j = 2$:

$$F_{12} = \frac{f_2^+ - f_{12}}{f_2^+ - f_2^-} = \frac{1.08 - 2.33}{1.08 - 8.88} = 0.1603$$

Likewise, calculations are carried out for all other values in the ACM. The complete results, composing the weighted ACM, are shown in Table 7.8. Notice that the range of each (normalized) criterion is from 0.0 to 1. Further, each normalized value in Table 7.8 = (1.0 – corresponding normalized value in Table 7.5).

TABLE 7.8

Normalized ACM for VIKOR Walk-Through.

Alternatives	C1	C2	C3
A1	1.0000	0.1603	0.5852
A2	0.8270	0.0000	0.8163
A3	0.5151	0.6885	1.0000
A4	0.2844	1.0000	0.0000
A5	0.0000	0.8090	0.4919

Step 3. Compute the group utility measure (S_i), which is the WS of normalized deviations of the ith alternative to the PIS (in Table 7.8) across all criteria. Further, compute the individual regret measure (R_i), which is the maximum normalized deviation of the ith alternative to

Where $a^{(2)}$ is the alternative ranked second in the ranking list based on Q. This condition ensures that the top-ranked alternative is significantly better than the second-best alternative.

Condition-2. Acceptable stability in decision-making, that is, the alternative $a^{(1)}$ must also be the best ranked by S and/or R. This condition ensures that the best-ranked alternative is stable under different evaluation perspectives, whether based on group benefit, individual regret, or a combination of both.

If one of the above conditions is not satisfied, then a set of compromise solutions is proposed, which is comprised of:

- Alternatives $a^{(1)}$ and $a^{(2)}$, if only Condition-2 is not satisfied, or
- Alternatives $a^{(1)}, a^{(2)}, \ldots, a^{(d)}$, if Condition-1 is not satisfied; here, $a^{(d)}$ is the dth-ranked alternative in Q and is determined by the following relation for the maximum d.

$$Q\left(\mathrm{a}^{(d)}\right) - Q\left(a^{(1)}\right) < \frac{1}{d-1} \tag{7.22}$$

NUMERICAL CALCULATIONS

Checking Condition-1: inspecting the column Q_i from Table 7.9, the second-ranked alternative $a^{(2)}$ is A2.

$$Q\left(a^{(2)}\right) - Q\left(a^{(1)}\right) = 0.1567 - 0.0845 = 0.0723$$

$$\frac{1}{m-1} = \frac{1}{5-1} = 0.25$$

Using Equation 7.21, since 0.0723 is not greater than or equal to 0.25, Condition-1 is not satisfied. Therefore, a set of compromise solutions is proposed, following Equation 7.22.

For instance, when $d = 3$, the third-ranked alternative $a^{(3)}$ is A5.

$$Q\left(\mathrm{a}^{(3)}\right) - Q\left(a^{(1)}\right) = 0.2888 - 0.0845 = 0.2044$$

$$\frac{1}{d-1} = \frac{1}{3-1} = 0.5$$

Since $0.2044 < 0.5$, A5 is included in the compromise solution set.

When $d = 4$, the fourth-ranked alternative $a^{(4)}$ is A4.

$$Q\left(a^{(4)}\right) - Q\left(a^{(1)}\right) = 0.5000 - 0.0845 = 0.4155$$

$$\frac{1}{d-1} = \frac{1}{4-1} = 0.3333$$

Since 0.4155 is not less than 0.3333, A4 is not included in the compromise solution set.

Finally, the complete set of compromise solutions in this example consists of the alternatives {A1, A2, A5}.

The advantages of VIKOR are as follows: (1) VIKOR uniquely balances overall distance from the best (i.e., group utility or benefit, S_i) and the worst single-criterion shortfall (i.e., individual regret, R_i). This suits problems where decision-makers would like to avoid terrible performance on any one criterion and keep overall performance high. (2) The parameter γ can be intuitively interpreted as the relative importance of group utility versus individual regret. Decision-makers can adjust it to reflect their specific attitudes (e.g., risk-averse might focus more on R_i and use a smaller γ value). (3) The last step of VIKOR can include conditions to ensure that the top-ranked alternative is a meaningful compromise. If the best alternative is not distinctly better than the second best, a tie or set of compromise solutions is recommended to decision-makers. (4) VIKOR has been extensively applied in various fields (Gul et al., 2016).

The limitations of VIKOR are as follows: (1) As with other reference-type methods that rely on PIS and/or NIS, adding or removing an alternative can cause rank reversal. (2) The parameter γ is somewhat subjective. A small shift in γ can possibly change the final ranking significantly. Without a clear rationale for choosing γ, decision-makers might question the final recommended results. (3) VIKOR does not always recommend a single top-ranked alternative but instead a set of compromise solutions in some cases (e.g., the ACM employed in this chapter). This requires an extra step from decision-makers to determine the final selection, which introduces more subjectivity. (4) VIKOR's compromise acceptability conditions and the interplay of S_i, R_i and Q_i can appear more complicated than other simpler MCDM methods, making it particularly challenging for users without a strong grasp of the underlying mathematics.

7.5 Evaluation Based on Distance from Average Solution (EDAS)

The EDAS method, introduced by Keshavarz Ghorabaee et al. (2015), evaluates alternatives by calculating their positive and negative distances from the average solution. Unlike methods such as TOPSIS or VIKOR that rely on PIS and NIS, EDAS is centered on how alternatives deviate from the average values across criteria. Positive distance from the average (PDA) reflects the extent to which an alternative exceeds the average, while negative distance from the average (NDA) measures the extent to which an alternative falls below the average. The assessment of alternatives is based on higher PDA values and lower NDA values, which indicate better performance relative to the average solution. Alternatives are ranked after computing their performance scores, which are derived by combining the normalized PDA and NDA values. The EDAS method can be performed in a structured stepwise manner, as follows:

Step 1. Determine the average solution, $\bar{A}$, of the original ACM with m rows (i.e., alternatives) and n columns (i.e., criteria).

$$\bar{f}_j = \frac{\sum_{i=1}^{m} f_{ij}}{m}$$
$$\bar{A} = \left\{\bar{f}_1, \bar{f}_2, \bar{f}_3, \ldots, \bar{f}_j, \ldots, \bar{f}_n\right\} \tag{7.23}$$

NUMERICAL CALCULATIONS (USING ACM FROM TABLE 7.1)

For instance, for $j = 1$:

$$\bar{f}_1 = \frac{\sum_{i=1}^{5} f_{i1}}{5} = \frac{0.185 + 0.317 + 0.555 + 0.731 + 0.948}{5} = 0.5472$$

Likewise, calculations are carried out for the other criteria. The average solution is found to be: $\bar{A} = \left\{\bar{f}_1, \bar{f}_2, \bar{f}_3\right\} = \{0.5472, 5.226, 458.4\}$

Step 2. Calculate the PDA and NDA matrices based on the determined average solution.

If the jth criterion is for maximization:

$$PDA_{ij} = \frac{max\left(0, \left(f_{ij} - \bar{f}_j\right)\right)}{\bar{f}_j}$$
$$NDA_{ij} = \frac{max\left(0, \left(\bar{f}_j - f_{ij}\right)\right)}{\bar{f}_j} \tag{7.24a}$$

If the jth criterion is for minimization:

$$PDA_{ij} = \frac{max\left(0,\left(\bar{f}_j - f_{ij}\right)\right)}{\bar{f}_j}$$
$$NDA_{ij} = \frac{max\left(0,\left(f_{ij} - \bar{f}_j\right)\right)}{\bar{f}_j} \tag{7.24b}$$

Here, PDA_{ij} and NDA_{ij} represent the positive distance (indicating better-than-average performance) and negative distance (indicating worse-than-average performance) of the ith alternative from the average solution with respect to the jth criterion, respectively.

NUMERICAL CALCULATIONS

For instance, for $i = 5$ and $j = 1$, since it is a maximization criterion:

$$PDA_{51} = \frac{max\left(0,\left(f_{51} - \bar{f}_1\right)\right)}{\bar{f}_1} = \frac{max(0,(0.948 - 0.5472))}{0.5472} = 0.7325$$

$$NDA_{51} = \frac{max\left(0,\left(\bar{f}_1 - f_{51}\right)\right)}{\bar{f}_1} = \frac{max(0,(0.5472 - 0.948))}{0.5472} = 0$$

For $i = 3$ and $j = 2$, since it is a minimization criterion:

$$PDA_{32} = \frac{max\left(0,\left(\bar{f}_2 - f_{32}\right)\right)}{\bar{f}_2} = \frac{max(0,(5.226 - 6.45))}{5.226} = 0$$

$$NDA_{32} = \frac{max\left(0,\left(f_{32} - \bar{f}_2\right)\right)}{\bar{f}_2} = \frac{max(0,(6.45 - 5.226))}{5.226} = 0.2342$$

Similar calculations are performed for all other values in the ACM and the complete PDA and NDA matrices are shown in Table 7.10. Note that, for an alternative and a criterion combination, either PDA or NDA is zero.

TABLE 7.10

PDA and NDA Matrices for EDAS Walk-Through.

	PDA			NDA		
Alternatives	**C1**	**C2**	**C3**	**C1**	**C2**	**C3**
A1	0	0.5542	0	0.6619	0	0.0096
A2	0	0.7933	0	0.4207	0	0.3499
A3	0.0143	0	0	0	0.2342	0.6204
A4	0.3359	0	0.8521	0	0.6992	0
A5	0.7325	0	0.1278	0	0.4141	0

Step 3. Calculate the WS of PDA and NDA (i.e., SP and SN) for each alternative.

$$SP_i = \sum_{j=1}^{n} w_j PDA_{ij} \tag{7.25}$$

$$SN_i = \sum_{j=1}^{n} w_j NDA_{ij} \tag{7.26}$$

NUMERICAL CALCULATIONS

For instance, for $i = 1$:

$$SP_1 = \sum_{j=1}^{3} w_j PDA_{1j} = 0.25 \times 0 + 0.4 \times 0.5542 + 0.35 \times 0 = 0.2217$$

$$SN_1 = \sum_{j=1}^{3} w_j NDA_{1j} = 0.25 \times 0.6619 + 0.4 \times 0 + 0.35 \times 0.0096 = 0.1688$$

Analogously, calculations are carried out for the remaining alternatives. The complete SP_i and SN_i results are shown in Table 7.11.

TABLE 7.11

Weighted Sum (SP_i and SN_i) and Normalized Weighted Sum (NSP_i and NSN_i) of PDA and NDA for EDAS Walk-Through.

Alternatives	SP_i	SN_i	NSP_i	NSN_i
A1	0.2217	0.1688	0.5800	0.4568
A2	0.3173	0.2276	0.8303	0.2676

TABLE 7.11 (*Continued*)

Weighted Sum (SP_i and SN_i) and Normalized Weighted Sum (NSP_i and NSN_i) of PDA and NDA for EDAS Walk-Through.

Alternatives	SP_i	SN_i	NSP_i	NSN_i
A3	0.0036	0.3108	0.0093	0.0000
A4	0.3822	0.2797	1.0000	0.1002
A5	0.2279	0.1656	0.5962	0.4671

Step 4. Normalize the values of SP_i and SN_i for each alternative.

$$NSP_i = \frac{SP_i}{\max_k (SP_k)} \qquad k \in \{1,2,\ldots,m\} \tag{7.27}$$

$$NSN_i = 1 - \frac{SN_i}{\max_k (SN_k)} \qquad k \in \{1,2,\ldots,m\} \tag{7.28}$$

When normalizing SN_i, the optimization directions of SN_i and NSN_i become opposite due to the introduction of a negative sign in the equation.

NUMERICAL CALCULATIONS

For instance, for $i = 1$:

$$NSP_1 = \frac{SP_1}{max_k (SP_k)} = \frac{0.2217}{0.3822} = 0.5800$$

$$NSN_1 = 1 - \frac{SN_1}{max_k (SN_k)} = 1 - \frac{0.1688}{0.3108} = 0.4568$$

The complete NSP_i and NSN_i results are presented in the last two columns in Table 7.11.

Step 5. Compute the performance score (P_i) of each alternative as follows:

$$P_i = \frac{1}{2}(NSP_i + NSN_i) \tag{7.29}$$

Finally, the alternative with the largest P_i is top-ranked and recommended to the decision-maker.

NUMERICAL CALCULATIONS

For instance, for $i = 1$:

$$P_1 = \frac{1}{2}(NSP_1 + NSN_1) = 0.5800 + 0.4568 = 0.5184$$

Similarly, the performance scores for all alternatives are calculated as $P_1 = 0.5184$, $P_2 = 0.5490$, $P_3 = 0.0047$, $P_4 = 0.5501$ and $P_5 = 0.5316$. Accordingly, the ranking is A4 > A2 > A5 > A1 > A3, with A4 being the top-ranked alternative by EDAS.

The advantages of EDAS are as follows: (1) It uses the average of each criterion as its reference solution, which is simple and intuitive. Decision-makers can readily interpret "positive distance" (better-than-average) and "negative distance" (worse-than-average). (2) By using the average solution, it is less affected by extreme values compared to methods that rely on ideal solutions. (3) It is computationally less intensive because of its simple calculations based on deviations from the average solution.

The limitations of EDAS are as follows: (1) Though less affected, adding or removing alternatives can still alter the average solution, potentially causing rank reversals. (2) Unlike some methods that explicitly compare each alternative to PIS and NIS, EDAS considers "above or below average". However, in some cases, decision-makers may need absolute benchmarks rather than relative-to-mean comparisons. (3) EDAS calculates the performance score as the average of NSP_i and NSN_i (i.e., equal weights by default). This aggregation might not always capture the true preferences of decision-makers.

7.6 Multi-Attributive Border Approximation Area Comparison (MABAC)

The MABAC method, introduced by Pamučar and Ćirović (2015), is based on the core principle of finding the performance score for each alternative by determining the sum of its distances from a unique reference solution, known as the border approximation area matrix. This matrix is made up of the geometric mean of each weighted normalized criterion, serving as a midway or baseline level of performance for comparison (Wang et al., 2022). The ranking is determined by evaluating how far each alternative deviates from this baseline, with a greater distance indicating a higher-ranked alternative

(Nabavi et al., 2023). The implementation of the MABAC method progresses through the following steps.

Step 1. Normalize the original ACM with m rows (i.e., alternatives) and n columns (i.e., criteria) using the max–min normalization method.

$$F_{ij} = \frac{f_{ij} - \min_{k\in\{1,2,\ldots,m\}} f_{kj}}{\max_{k\in\{1,2,\ldots,m\}} f_{kj} - \min_{k\in\{1,2,\ldots,m\}} f_{kj}} \quad \text{for maximization criterion} \tag{7.30}$$

$$F_{ij} = \frac{\max_{k\in\{1,2,\ldots,m\}} f_{kj} - f_{ij}}{\max_{k\in\{1,2,\ldots,m\}} f_{kj} - \min_{k\in\{1,2,\ldots,m\}} f_{kj}} \quad \text{for minimization criterion} \tag{7.31}$$

Notably, the above normalization equation for a minimization criterion reverses its optimization direction, changing from minimizing f_{ij} to maximizing F_{ij} because of the introduction of a negative sign to f_{ij} in the numerator. Additionally, all values of F_{ij} are in the range of [0, 1].

NUMERICAL CALCULATIONS (USING ACM FROM TABLE 7.1)

For instance, for $i = 2$ and $j = 1$:

$$F_{21} = \frac{0.317 - 0.185}{0.948 - 0.185} = 0.1730$$

Similar calculations are performed for all other values in the ACM and the complete normalized ACM results are presented in Table 7.12.

TABLE 7.12

Normalized ACM for MABAC Walk-Through.

Alternatives	C1	C2	C3
A1	0.0000	0.8397	0.4148
A2	0.1730	1.0000	0.1837
A3	0.4849	0.3115	0.0000
A4	0.7156	0.0000	1.0000
A5	1.0000	0.1910	0.5081

Step 2. Construct the weighted normalized ACM using the assigned weights, w_j.

$$v_{ij} = \left(1 + F_{ij}\right) \times w_j \tag{7.32}$$

Note that a constant value of 1 is added to all F_{ij} values because the max–min normalization method inherently creates a value of 0 in each normalized criterion. This would cause the multiplication to become 0 in the subsequent step of calculating the geometric mean values. To prevent this, a constant value of 1 is introduced, as in Pamučar and Ćirović (2015).

NUMERICAL CALCULATIONS

For instance, for $i = 2$ and $j = 1$:

$$v_{21} = (1 + F_{21}) \times w_1 = (1 + 0.1730) \times 0.25 = 0.2933$$

Likewise, calculations are carried out for all other values in the normalized ACM, using their respective assigned weights. The results constituting the weighted normalized ACM are presented in Table 7.13.

TABLE 7.13

Weighted Normalized ACM for MABAC Walk-Through.

Alternatives	C1	C2	C3
A1	0.2500	0.7359	0.4952
A2	0.2933	0.8000	0.4143
A3	0.3712	0.5246	0.3500
A4	0.4289	0.4000	0.7000
A5	0.5000	0.4764	0.5279

Step 3. Determine the border approximation area matrix B. Essentially, b_j sets the benchmark above which an alternative is comparatively good (for that criterion) and below which it is comparatively poor.

$$b_j = \left(\prod_{i=1}^{m} v_{i1} \right)^{1/m} \tag{7.33}$$

$$B = \{b_1, b_2, \ldots, b_j, \ldots, b_n\}$$

Note that the border approximation area matrix B is a row vector with n columns.

NUMERICAL CALCULATIONS

For instance, for $j = 1$:

$$b_1 = \left(\prod_{i=1}^{5} v_{i1}\right)^{1/5} = (0.2500 \times 0.2933 \times 0.3712 \times 0.4289 \times 0.5000)^{1/5} = 0.3575$$

Likewise, the border approximation area matrix B is found to be:

$$B = \{b_1, b_2, b_3\} = \{0.3575, 0.5675, 0.4839\}$$

Step 4. Compute the performance score (P_i) of each alternative.

$$P_i = \sum_{j=1}^{n} (v_{ij} - b_j) \tag{7.34}$$

The alternative with the largest P_i is top-ranked and recommended to the decision-maker.

NUMERICAL CALCULATIONS

For instance, for $i = 2$: $P_2 = \sum_{j=1}^{3} (v_{2j} - b_j)$

$$= (0.2933 - 0.3575) + (0.8000 - 0.5675) + (0.4143 - 0.4839) = 0.0987$$

Similarly, the performance scores for all alternatives are calculated as $P_1 = 0.0722$, $P_2 = 0.0987$, $P_3 = -0.1630$, $P_4 = 0.1201$ and $P_5 = 0.0954$. Accordingly, the ranking is A4 > A2 > A5 > A1 > A3, with A4 being the top-ranked alternative by MABAC.

The advantages of MABAC are as follows: (1) Its novelty lies in utilizing the border approximation area matrix, derived from the geometric mean, as the reference solution. Alternatives are evaluated based on whether they lie above or below this border across the criteria. (2) The computational procedure is straightforward; after normalization and weighting, the position of each alternative relative to the border is determined through a simple (matrix) subtraction. (3) Unlike methods that rely on PIS and/or NIS, MABAC assesses alternatives relative to the geometric mean-based border approximation, making it less susceptible to the extreme values.

The limitations of MABAC are as follows: (1) The concept of the "border" may be abstract for decision-makers accustomed to explicit PIS and/or NIS as reference points. In other words, comparing alternatives to an intermediate baseline may feel less intuitive. (2) Although MABAC is less sensitive to extreme values than some methods, outliers can still distort the geometric mean-based border approximation area matrix, affecting all subsequent calculations. (3) Also, it suffers from rank reversal in the case of the addition or removal of alternatives.

7.7 Combinative Distance-Based Assessment (CODAS)

The CODAS method, developed by Keshavarz Ghorabaee et al. (2016), is underpinned by the core principle of finding the performance score for each alternative by calculating both Euclidean and Taxicab distances from the NIS reference point. The alternative that is farther from the NIS is ranked higher (Wang et al., 2024). If there are two alternatives that are incomparable according to Euclidean distance, then the Taxicab distance is used as the secondary measure. The steps in the CODAS method are as follows:

Step 1. Normalize the original ACM with m rows (i.e., alternatives) and n columns (i.e., criteria) using the max normalization method (as in Keshavarz Ghorabaee et al., 2016).

$$F_{ij} = \frac{f_{ij}}{\max_{k \in \{1,2,\ldots,m\}} f_{kj}} \quad \text{for maximization criterion} \tag{7.35}$$

$$F_{ij} = \frac{\min_{k \in \{1,2,\ldots,m\}} f_{kj}}{f_{ij}} \quad \text{for minimization criterion} \tag{7.36}$$

The above normalization equation for a minimization criterion alters the optimization direction of the criterion from minimizing f_{ij} (i.e., the smaller f_{ij}, the better) to maximizing F_{ij} (i.e., the larger F_{ij}, the better), as f_{ij} is positioned in the denominator of the equation.

NUMERICAL CALCULATIONS (USING ACM FROM TABLE 7.1)

For instance, for $i = 2$ and $j = 1$:

$$F_{21} = \frac{0.317}{0.948} = 0.3344$$

Similar calculations are performed for all other values in the ACM and the complete normalized ACM results are presented in Table 7.14.

TABLE 7.14

Normalized ACM for the CODAS Walk-Through.

Alternatives	C1	C2	C3
A1	0.1951	0.4635	0.5347
A2	0.3344	1.0000	0.3510
A3	0.5854	0.1674	0.2049
A4	0.7711	0.1216	1.0000
A5	1.0000	0.1461	0.6090

Step 2. Construct the weighted normalized ACM by multiplying the values of each criterion by its assigned weight, w_j.

$$v_{ij} = F_{ij} \times w_j \tag{7.37}$$

NUMERICAL CALCULATIONS

For instance, for $i = 2$ and $j = 1$:

$$v_{21} = F_{21} \times w_1 = 0.3344 \times 0.25 = 0.0836$$

Likewise, calculations are carried out for all other values in the normalized ACM, using their respective assigned weights. The resulting weighted normalized ACM is presented in Table 7.15.

TABLE 7.15

Weighted Normalized ACM for the CODAS Walk-Through.

Alternatives	C1	C2	C3
A1	0.0488	0.1854	0.1872
A2	0.0836	0.4000	0.1229
A3	0.1464	0.0670	0.0717
A4	0.1928	0.0486	0.3500
A5	0.2500	0.0585	0.2131

Step 3. Determine the NIS, A^-, from the weighted normalized ACM.

$$\begin{aligned} A^- &= \left\{\min_i\left(v_{ij}\right)\right\} \\ &= \left\{v_1^-, v_2^-, \ldots, v_j^-, \ldots, v_n^-\right\} \end{aligned} \tag{7.38}$$

The above equation to determine A^- differs from that of TOPSIS due to the application of the max normalization method, which transforms minimization criteria into maximization type.

NUMERICAL CALCULATIONS

For instance, for $j = 1$:

$$v_1^- = \min_i\left(v_{i1}\right) = 0.0488$$

The NIS is found to be: $A^- = \left\{v_1^-, v_2^-, v_3^-\right\} = \{0.0488, 0.0486, 0.0717\}$, which can be verified by reviewing the weighted normalized values in Table 7.15.

Step 4. Compute the Euclidean and Taxicab distances of each alternative to the NIS.

$$E_i = \sqrt{\sum_{j=1}^{n}\left(v_{ij} - v_j^-\right)^2} \tag{7.39}$$

$$T_i = \sum_{j=1}^{n}\left|v_{ij} - v_j^-\right| \tag{7.40}$$

NUMERICAL CALCULATIONS

For instance, for $i = 1$:

$$\begin{aligned} E_1 &= \sqrt{\sum_{j=1}^{3}\left(v_{1j} - v_j^-\right)^2} = \sqrt{\begin{aligned}&(0.0488 - 0.0488)^2 + (0.1854 - 0.0486)^2 \\ &+ (0.1872 - 0.0717)^2\end{aligned}} \\ &= 0.1790 \end{aligned}$$

$$T_1 = \sum_{j=1}^{3} \left| v_{1j} - v_j^- \right| = \left|0.0488 - 0.0488\right| + \left|0.1854 - 0.0486\right| +$$
$$\left|0.1872 - 0.0717\right| = 0.2522$$

Analogously, it can be computed that $E_2 = 0.3568$ and $T_2 = 0.4373$ for $i = 2$. The complete Euclidean and Taxicab distances are presented in Table 7.16.

TABLE 7.16

Euclidean and Taxicab Distances of Each Alternative to the Negative-Ideal Solutions for the CODAS Walk-Through.

Alternatives	Euclidean Distance	Taxicab Distance
A1	0.1790	0.2522
A2	0.3568	0.4373
A3	0.0993	0.1159
A4	0.3133	0.4223
A5	0.2461	0.3524

Step 5. Construct the relative assessment matrix H.

$$h_{ik} = (E_i - E_k) + \psi(E_i - E_k) \times (T_i - T_k) \qquad i,k \in \{1,2,\ldots,m\} \tag{7.41}$$

Here, $\in [0.01, 0.05]$ and is selected to be 0.02 (as in Keshavarz Ghorabaeeetal.,2016).Furthermore, $\psi(E_i - E_k) = 1$ if $|E_i - E_k| \geq \tau$, and $= 0$ if $|E_i - E_k| < \tau$.

NUMERICAL CALCULATIONS

For instance, for $i = 2$ and $k = 1$:

$$|E_2 - E_1| = |0.3568 - 0.1790| = 0.1778 \geq 0.02$$

$$\text{and so } \psi(E_2 - E_1) = 1$$

Then,

$$\begin{aligned} h_{21} &= (E_2 - E_1) + \psi(E_2 - E_1) \times (T_2 - T_1) \\ &= (0.3568 - 0.1790) + 1 \times (0.4373 - 0.2522) = 0.3629 \end{aligned}$$

Likewise, other values in the H matrix can be computed. The complete H matrix is presented in Table 7.17.

TABLE 7.17

H Matrix for the CODAS Walk-Through.

Alternatives	A1	A2	A3	A4	A5
A1	0	−0.3629	0.2160	−0.3044	−0.1674
A2	0.3629	0	0.5789	0.0585	0.1955
A3	−0.2160	−0.5789	0	−0.5204	−0.3834
A4	0.3044	−0.0585	0.5204	0	0.1370
A5	0.1674	−0.1955	0.3834	−0.1370	0

Step 6. Compute the performance score (P_i) of each alternative as follows:

$$P_i = \sum_{k=1}^{m} h_{ik} \tag{7.42}$$

The alternative with the largest P_i is top-ranked and recommended to the decision-maker.

NUMERICAL CALCULATIONS

For instance, for $i = 2$:

$$P_2 = \sum_{k=1}^{5} h_{2k} = 0.3629 + 0 + 0.5789 + 0.0585 + 0.1955 = 1.1957$$

Similarly, performance scores for all alternatives are calculated as $P_1 = -0.6187$, $P_2 = 1.1957$, $P_3 = -1.6986$, $P_4 = 0.9034$ and $P_5 = 0.2183$. Accordingly, the ranking is A2 > A4 > A5 > A1 > A3, with A2 being the top-ranked alternative. As can be seen, P_i can be negative.

The advantages of CODAS are as follows: (1) The use of the max normalization method effectively transforms minimization criteria into maximization criteria, which slightly simplifies the subsequent calculations in determining the NIS. (2) It uses both Euclidean and Taxicab distances, improving the ability to distinguish alternatives and providing a more nuanced ranking.

(3) The introduction of the novel threshold parameter, τ, allows for an integrated use of Euclidean and Taxicab distances, theoretically enabling greater flexibility in addressing complex decision-making scenarios.

The limitations of CODAS are as follows: (1) Choosing an appropriate τ is nontrivial and may require input from the decision-maker; an improper τ value can cause erratic or inconsistent rankings. (2) CODAS references only the NIS, which might not align with the preferences of decision-makers who favor anchoring their evaluations to the PIS as well. (3) In the original CODAS method, the threshold function ψ is defined as $\psi(E_i - E_k) = 1 \text{ if} |E_i - E_k| \geq \tau, \text{and} = 0 \text{if} |E_i - E_k| < \tau$. However, Chen and Gou (2022) pointed out that this approach is inconsistent with realistic decision-making. They stated that, when the Euclidean distance difference between two alternatives is tiny (i.e., less than the threshold), the Taxicab distance should be incorporated to provide additional differentiation. On the contrary, when the Euclidean distance difference between two alternatives is large (i.e., greater than or equal to the threshold), the Euclidean distance alone should suffice to distinguish between the two alternatives, making the use of Taxicab distance unnecessary.

7.8 Proximity Indexed Value (PIV)

The PIV method, introduced by Mufazzal and Muzakkir (2018) to mitigate the rank reversal phenomenon, is based on the core principle of determining the performance score of each alternative by comparing it against the PIS. Alternatives that are closer to the PIS receive higher rankings. The weighted proximity index is determined by calculating the difference between the weighted normalized value (v_{ij}) and the best value of each criterion. The PIV method is implemented through the following series of steps.

Step 1. Normalize the original ACM with m rows (i.e., alternatives) and n columns (i.e., criteria) using the vector normalization method. This step does not differ from Step 1 of the TOPSIS method and Equation 7.1 is equally applicable here. For brevity, this equation and numerical calculations (same as Table 7.2) are not repeated here.

Step 2. Construct the weighted normalized ACM using the assigned weights, w_j. As with Step 1, Equation 7.2 is also applicable here. The equation and numerical calculations (same as Table 7.3) are omitted here.

Step 3. Determine the best value for each criterion (v_j^+), forming the PIS. This step is the same as Step 3 of the TOPSIS method and utilizes Equation 7.3. The PIS is also found to be: $A^+ = \{v_1^+, v_2^+, v_3^+\} = \{0.1730, 0.0321, 0.2593\}$.

Step 4. Calculate the overall proximity value (D_i) of each alternative to the PIS by:

$$D_i = \sum_{j=1}^{n} \left| v_{ij} - v_j^{+} \right| \tag{7.43}$$

The alternative with the smallest D_i is top-ranked and recommended to the decision-maker. In effect, the PIV method finds the alternative closest to the PIS (as measured by D_i).

NUMERICAL CALCULATIONS

For instance, for $i = 1$:

$$D_1 = \sum_{j=1}^{3} \left| v_{1j} - v_j^{+} \right| = |0.0338 - 0.1730| + |0.0691 - 0.0321| +$$
$$|0.1387 - 0.2593| = 0.2970$$

Similarly, the overall proximity values for all alternatives are calculated as $D_1 = 0.2970$, $D_2 = 0.2835$, $D_3 = 0.4373$, $D_4 = 0.2711$ and $D_5 = 0.2887$. Accordingly, the ranking is A4 ≻ A2 ≻ A5 ≻ A1 ≻ A3, with A4 being the top-ranked alternative by PIV.

The advantages of PIV are as follows: (1) It focuses on the distance from the PIS only, reducing conceptual (and computational) overhead compared to other reference-type methods. (2) The essential usage of the L1 norm (i.e., summation of absolute deviations from the PIS) is intuitive for decision-makers. (3) The concept and implementation of PIV are easy to understand.

The limitations of PIV are as follows: (1) By not considering the NIS, it fails to capture how far alternatives are from the worst-case scenario, which could be a critical consideration in risk-averse applications. (2) It is more susceptible to the influence of extreme values. If the identified reference solution is an extreme outlier, it can skew all distances significantly. (3) L1 norm treats all deviations linearly with equal emphasis, potentially under-penalizing a big shortfall in one criterion. (4) The default use of the Max normalization method may result in a division by zero error.

7.9 Measurement of Alternatives and Ranking According to Compromise Solution (MARCOS)

The MARCOS method, introduced by Stević et al. (2020), is based on the principle of measuring alternatives and ranking them according to their compromise with both the PIS and NIS. By establishing the relationship of each alternative to these two reference points, MARCOS determines utility functions and their aggregation, which reflect the proximity of an alternative to the PIS and its distance from the NIS. An alternative is ranked higher if it is closer to PIS and farther from NIS. According to Stević et al. (2020), this method introduces a novel approach to defining and aggregating utility functions, enhancing the decision-making framework. They also performed comprehensive sensitivity analyses, including variations in weight values, decision-making matrices and comparisons with other MCDM methods, demonstrating its robustness and validity across different scenarios. The MARCOS method operates through the following step-by-step procedure.

Step 1. Define the PIS, A^+ and the NIS, A^-:

$$\begin{aligned} A^+ &= \left\{ \left(\max_i \left(f_{ij} \right) | j \in J_{max} \right), \left(\min_i \left(f_{ij} \right) | j \in J_{min} \right) \right\} \\ &= \left\{ f_1^+, f_2^+, \ldots, f_j^+, \ldots, f_n^+ \right\} \end{aligned} \tag{7.44}$$

$$\begin{aligned} A^- &= \left\{ \left(\min_i \left(f_{ij} \right) | j \in J_{max} \right), \left(\max_i \left(f_{ij} \right) | j \in J_{min} \right) \right\} \\ &= \left\{ f_1^-, f_2^-, \ldots, f_j^-, \ldots, f_n^- \right\} \end{aligned} \tag{7.45}$$

In both the above equations, J_{max} is the set of maximization criteria and J_{min} is the set of minimization criteria, both from $\{1,2,\ldots,n\}$. Then, add both the PIS and NIS to the original ACM (with m alternatives and n criteria), forming an extended ACM, as illustrated below.

NUMERICAL CALCULATIONS (USING ACM FROM TABLE 7.1)

For instance, for $j=1$:

$$f_1^+ = \max_i \left(f_{i1} \right) = 0.948$$

$$f_1^- = \min_i \left(f_{i1} \right) = 0.185$$

For $j=2$:

$$f_2^+ = \min_i \left(f_{i2} \right) = 1.08$$

$$f_2^- = \max_i \left(f_{i2} \right) = 8.88$$

The PIS is found to be: $A^+ = \left\{ f_1^+, f_2^+, f_3^+ \right\} = \{0.948, 1.08, 849\}$. The NIS is found to be: $A^- = \left\{ f_1^-, f_2^-, f_3^- \right\} = \{0.185, 8.88, 174\}$. The extended ACM is shown in Table 7.18, with A^+ and A^- appended as the last two rows.

TABLE 7.18

Extended ACM for MARCOS Walk-Through.

Alternatives	C1	C2	C3
A1	0.185	2.33	454
A2	0.317	1.08	298
A3	0.555	6.45	174
A4	0.731	8.88	849
A5	0.948	7.39	517
A^+	0.948	1.08	849
A^-	0.185	8.88	174

Step 2. Normalize the extended ACM (with $i \in \{1,2,\ldots,m+2\}$) using the max normalization method (as in Stević et al., 2020).

$$F_{ij} = \frac{f_{ij}}{\max\limits_{k\in\{1,2,\ldots,m\}} f_{kj}} \quad \text{for maximization criterion} \tag{7.46}$$

$$F_{ij} = \frac{\min\limits_{k\in\{1,2,\ldots,m\}} f_{kj}}{f_{ij}} \quad \text{for minimization criterion} \tag{7.47}$$

Use of the above normalization equation for a minimization criterion alters the optimization direction of the criterion from minimizing f_{ij} (i.e., the smaller f_{ij}, the better) to maximizing F_{ij} (i.e., the larger F_{ij}, the better), as f_{ij} is in the denominator of the equation.

NUMERICAL CALCULATIONS

For instance, for $i = 2$ and $j = 1$:

$$F_{21} = \frac{0.317}{0.948} = 0.3344$$

Similar calculations are performed for all other values in the extended ACM and the complete normalized extended ACM results are presented in Table 7.19.

TABLE 7.19

Normalized Extended ACM for MARCOS Walk-Through.

Alternatives	C1	C2	C3
A1	0.1951	0.4635	0.5347
A2	0.3344	1.0000	0.3510
A3	0.5854	0.1674	0.2049
A4	0.7711	0.1216	1.0000
A5	1.0000	0.1461	0.6090
A^+	1.0000	1.0000	1.0000
A^-	0.1951	0.1216	0.2049

Step 3. Construct the weighted normalized extended ACM (with $i \in \{1,2,\ldots,m+2\}$) by multiplying the values of each criterion by its assigned weight, w_j.

$$v_{ij} = F_{ij} \times w_j \tag{7.48}$$

As seen from the first three steps, MARCOS initially identifies the ideal solutions, followed by normalization and then applying weight. This sequence essentially leads to the same outcome as the conventional approach of normalizing first, then applying weights and subsequently determining ideal solutions. This can be observed by comparing the following Table 7.20 with Table 7.15 in the CODAS section (as both MARCOS and CODAS employ the max normalization method).

NUMERICAL CALCULATIONS

For instance, for $i = 2$ and $j = 1$:

$$v_{21} = F_{21} \times w_1 = 0.3344 \times 0.25 = 0.0836$$

Likewise, calculations are carried out for all other values in the normalized extended ACM, using their respective assigned weights. The results, which constitute the weighted normalized extended ACM, are presented in Table 7.20.

TABLE 7.20

Weighted Normalized Extended ACM for MARCOS Walk-Through.

Alternatives	C1	C2	C3
A1	0.0488	0.1854	0.1872
A2	0.0836	0.4000	0.1229
A3	0.1464	0.0670	0.0717
A4	0.1928	0.0486	0.3500
A5	0.2500	0.0585	0.2131
A^+	0.2500	0.4000	0.3500
A^-	0.0488	0.0486	0.0717

Step 4. Calculate the utility degrees (K_i^+ and K_i^-) and utility function values ($f\left(K_i^+\right)$ and $f\left(K_i^-\right)$) of alternatives in relation to the PIS and NIS, respectively, by the following equations.

$$S_i = \sum_{j=1}^{n} v_{ij} \qquad \text{for } i \in \{1,2,\ldots,m\} \tag{7.49a}$$

$$S_{A^+} = \sum_{j=1}^{n} v_{ij} \qquad \text{for } i = m+1 \tag{7.49b}$$

$$S_{A^-} = \sum_{j=1}^{n} v_{ij} \qquad \text{for } i = m+2 \tag{7.49c}$$

$$K_i^+ = \frac{S_i}{S_{A^+}} \qquad \text{for } i \in \{1,2,\ldots,m\} \tag{7.50}$$

$$K_i^- = \frac{S_i}{S_{A^-}} \qquad \text{for } i \in \{1,2,\ldots,m\} \tag{7.51}$$

Here, S_{A^+} and S_{A^-} represent the total sum of the weighted normalized values across all criteria for the PIS and NIS (A^+ and A^-), respectively.

$$f\left(K_i^+\right) = \frac{K_i^-}{K_i^+ + K_i^-} \qquad \text{for } i \in \{1,2,\ldots,m\} \tag{7.52}$$

$$f\left(K_i^-\right) = \frac{K_i^+}{K_i^+ + K_i^-} \qquad \text{for } i \in \{1,2,\ldots,m\} \tag{7.53}$$

NUMERICAL CALCULATIONS

$$S_{A^+} = 0.2500 + 0.4000 + 0.3500 = 1.0000$$

Note that the above quantity will always be unity due to max normalization and weights chosen to sum to unity.

$$S_{A^-} = 0.0488 + 0.0486 + 0.0717 = 0.1692$$

For instance, for $i = 1$:

$$S_1 = 0.0488 + 0.1854 + 0.1872 = 0.4214$$

$$K_1^+ = \frac{S_1}{S_{A^+}} = \frac{0.4214}{1} = 0.4214$$

$$K_1^- = \frac{S_1}{S_{A^-}} = \frac{0.4214}{0.1692} = 2.4908$$

$$f\left(K_1^+\right) = \frac{K_1^-}{K_1^+ + K_1^-} = \frac{2.4908}{0.4214 + 2.4908} = 0.8553$$

$$f\left(K_1^-\right) = \frac{K_1^+}{K_1^+ + K_1^-} = \frac{0.4214}{0.4214 + 2.4908} = 0.1447$$

Analogously, utility degrees (K_i^+ and K_i^-) and utility function values ($f\left(K_i^+\right)$ and $f\left(K_i^-\right)$) of the other alternatives are calculated and presented in Table 7.21. Notably, $f\left(K_i^+\right)$ and $f\left(K_i^-\right)$ values are identical for all alternatives because $f\left(K_i^+\right)$ equation can be simplified as follows:

$$f\left(K_i^+\right) = \frac{K_i^-}{K_i^+ + K_i^-} = \frac{\dfrac{S_i}{S_{A^-}}}{\dfrac{S_i}{S_{A^+}} + \dfrac{S_i}{S_{A^-}}} = \frac{\dfrac{1}{S_{A^-}}}{\dfrac{1}{S_{A^+}} + \dfrac{1}{S_{A^-}}}$$

This simplified expression depends solely on S_{A^+} and S_{A^-}, making utility function calculations invariant to the specific alternative.

TABLE 7.21

Utility Degrees and Utility Function Values for MARCOS Walk-Through.

Alternatives	S_i	K_i^+	K_i^-	$f(K_i^+)$	$f(K_i^-)$
A1	0.4214	0.4214	2.4908	0.8553	0.1447
A2	0.6064	0.6064	3.5849	0.8553	0.1447
A3	0.2851	0.2851	1.6851	0.8553	0.1447
A4	0.5914	0.5914	3.4961	0.8553	0.1447
A5	0.5216	0.5216	3.0833	0.8553	0.1447
A^+	1.0000				
A^-	0.1692				

Step 5. Compute the performance score (P_i) of each alternative as follows:

$$P_i = \frac{K_i^+ + K_i^-}{1 + \frac{1 - f(K_i^+)}{f(K_i^+)} + \frac{1 - f(K_i^-)}{f(K_i^-)}} \quad \text{for } i \in \{1, 2, \ldots, m\} \tag{7.54}$$

The alternative with the largest P_i is top-ranked and recommended to the decision-maker.

NUMERICAL CALCULATIONS

For instance, for $i = 1$:

$$P_1 = \frac{K_1^+ + K_1^-}{1 + \frac{1 - f(K_1^+)}{f(K_1^+)} + \frac{1 - f(K_1^-)}{f(K_1^-)}} = \frac{0.4214 + 2.4908}{1 + \frac{1 - 0.8553}{0.8553} + \frac{1 - 0.1447}{0.1447}} = 0.4113$$

Similarly, performance scores for all alternatives are calculated as $P_1 = 0.4113$, $P_2 = 0.5920$, $P_3 = 0.2783$, $P_4 = 0.5773$ and $P_5 = 0.5091$. Accordingly, the ranking is A2 ≻ A4 ≻ A5 ≻ A1 ≻ A3, with A2 being the top-ranked alternative.

The advantages of MARCOS are as follows: (1) By explicitly adding the PIS and NIS to the ACM matrix, the process becomes more transparent, allowing decision-makers to see exactly what (raw) reference points are being used as benchmarks. (2) The method has been successfully applied in diverse fields, such as supplier selection, health care, logistics and manufacturing, showcasing its good adaptability (Demir et al., 2024). (3) It delivers a single, intuitively interpretable indicator, that is, the utility degree, for every alternative. This scalar is computed from the alternatives' simultaneous proximity to the PIS and NIS, giving decision-makers an at-a-glance sense of "how good" each option is while still preserving full mathematical rigor.

The limitations of MARCOS are as follows: (1) Compared to simpler methods (e.g., EDAS), MARCOS involves more calculations, particularly in the utility function aggregation step, making it more computationally intensive when handling a large ACM. (2) Decision-makers unfamiliar with utility functions may find it challenging to interpret results without a solid understanding of the underlying mathematics. (3) Like other reference-type methods, MARCOS can exhibit rank reversal when the set of alternatives changes.

7.10 Preference Ranking on the Basis of Ideal-Average Distance (PROBID)

The PROBID method, proposed by Wang et al. (2021), is based on the core principle of determining the performance score of each alternative by comparing it against a range of reference solutions. This range encompasses a hierarchy of ideal solutions, starting from the most PIS and extending through lesser tiers (e.g., second and third most PIS) down to the mth PIS (i.e., the most NIS). Additionally, the PROBID method incorporates an average solution to provide a comprehensive baseline for comparison. Hence, according to Wang et al. (2021), this method is less susceptible to minor data changes, reducing the likelihood of rank reversal. By considering multiple layers of ideal solutions and the average solution, PROBID provides a more refined ranking of alternatives (Baydaş et al., 2024). Besides, a simplified variant, sPROBID, is also available in the literature (Nabavi et al., 2025). In PROBID, the alternative that is closer to the PIS group, farther from the NIS group and deviates more from the average solutions is ranked higher. The procedure of the PROBID method unfolds in a series of steps as follows:

Step 1. Normalize the original ACM with m rows (i.e., alternatives) and n columns (i.e., criteria) using the vector normalization method. This step does not differ from Step 1 of the TOPSIS method and Equation 7.1 is equally applicable here. For brevity, this equation and numerical calculations (same as Table 7.2) are not repeated here.

Step 2. Construct the weighted normalized ACM using the assigned weights, w_j. As with Step 1, Equation 7.2 is also applicable here. This equation and numerical calculations (same as Table 7.3) are omitted here.

Step 3. Determine the most PIS ($A_{(1)}$), second PIS ($A_{(2)}$), third PIS ($A_{(3)}$), . . . and m^{th} PIS ($A_{(m)}$) (i.e., the most NIS) by:

$$\begin{aligned} A_{(k)} &= \left\{ \left(\underset{i}{Large}\left(v_{ij},k\right) \mid j \in J_{max} \right), \left(\underset{i}{Small}\left(v_{ij},k\right) \mid j \in J_{min} \right) \right\} \\ &= \left\{ v_{(k)1}, v_{(k)2}, v_{(k)3}, \dots, v_{(k)j}, \dots, v_{(k)n} \right\} \end{aligned} \tag{7.55}$$

Here, J_{max} is the set of maximization criteria and J_{min} is the set of minimization criteria; $k \in \{1,2,\dots,m\}$; and $\underset{i}{Large}\left(v_{ij},k\right)$ is the kth largest value in the jth criterion (i.e., $v_{(k)j}$). Then, compute the average solution, $\bar{A}$ as follows:

$$\begin{aligned} \bar{v}_j &= \frac{\sum_{i=1}^{m} v_{ij}}{m} \\ \bar{A} &= \left\{ \bar{v}_1, \bar{v}_2, \bar{v}_3, \dots, \bar{v}_j, \dots, \bar{v}_n \right\} \end{aligned} \tag{7.56}$$

NUMERICAL CALCULATIONS

For instance, for $k = 2$, that is, finding the second best value of each criterion:

$$A_{(2)} = \left\{ v_{(2)1}, v_{(2)2}, v_{(2)3} \right\} = \{0.1334, 0.0691, 0.1579\}$$

Likewise, the other tiers of ideal solutions and the average solution can be determined. The complete results are shown in Table 7.22.

TABLE 7.22

Tiers of Ideal Solutions and the Average Solution for the PROBID Walk-Through.

Ideal Solutions	**C1**	**C2**	**C3**
$A_{(1)}$	0.1730	0.0321	0.2593
$A_{(2)}$	0.1334	0.0691	0.1579
$A_{(3)}$	0.1013	0.1914	0.1387
$A_{(4)}$	0.0579	0.2193	0.0910
$A_{(5)}$	0.0338	0.2635	0.0531
Average	0.0999	0.1551	0.1400

Step 4. Iteratively calculate the Euclidean distance of each alternative to each of the m ideal solutions as well as to the average solution.

$$\begin{aligned} &\textit{Distance to the } k^{th} \textit{ PIS}, S_{i(k)} = \sqrt{\sum_{j=1}^{n}\left(v_{ij} - v_{(k)j}\right)^2} \\ &\textit{Distance to the average solution}, S_{i(avg)} = \sqrt{\sum_{j=1}^{n}\left(v_{ij} - \overline{v}_j\right)^2} \end{aligned} \quad (7.57)$$

NUMERICAL CALCULATIONS

For instance, for $i = 1$ and $k = 2$:

$$\begin{aligned} S_{1(2)} &= \sqrt{\sum_{j=1}^{3}\left(v_{1j} - v_{(2)j}\right)^2} \\ &= \sqrt{(0.0338 - 0.1334)^2 + (0.0691 - 0.0691)^2 + (0.1387 - 0.1579)^2} \\ &= 0.1015 \end{aligned}$$

The complete distances of all alternatives are presented in Table 7.23. Note that the $S_{i(1)}$ and $S_{i(5)}$ are the same as the S_{i+} and S_{i-} of TOPSIS (in Table 7.4), respectively.

TABLE 7.23

Distances of Each Alternative to the Different Ideal Solutions and the Average Solution for the PROBID Walk-Through.

Alternatives	$S_{i(1)}$	$S_{i(2)}$	$S_{i(3)}$	$S_{i(4)}$	$S_{i(5)}$	$S_{i(avg)}$
A1	0.1880	0.1015	0.1397	0.1594	0.2124	0.1084
A2	0.2039	0.1075	0.1719	0.1873	0.2358	0.1389
A3	0.2703	0.1642	0.0855	0.0640	0.0988	0.0942
A4	0.2348	0.2192	0.1442	0.1897	0.2290	0.1647
A5	0.2130	0.1553	0.0793	0.1332	0.1798	0.0990

Step 5. Determine the overall positive-ideal and negative-ideal distances

$$S_{i(pos-ideal)} = \begin{cases} \sum_{k=1}^{\frac{m+1}{2}} \frac{1}{k} S_{i(k)} & \textit{when m is an odd number} \\ \sum_{k=1}^{\frac{m}{2}} \frac{1}{k} S_{i(k)} & \textit{when m is an even number} \end{cases}$$

$$S_{i(neg-ideal)} = \begin{cases} \sum_{k=\frac{m+1}{2}}^{m} \frac{1}{m-k+1} S_{i(k)} \text{ when } m \text{ is an odd number} \\ \sum_{k=\frac{m}{2}+1}^{m} \frac{1}{m-k+1} S_{i(k)} \text{ when } m \text{ is an even number} \end{cases} \tag{7.58}$$

Note that, when m is an odd number, $S_{i\left(\frac{m+1}{2}\right)}$ is used for calculating both $S_{i(pos-ideal)}$ and $S_{i(ne-ideal)}$.

NUMERICAL CALCULATIONS

For instance, for $i = 1$:

$$S_{1(pos-ideal)} = \sum_{k=1}^{3} \frac{1}{k} S_{1(k)} = 1 \times 0.1880 + \frac{1}{2} \times 0.1015 + \frac{1}{3} \times 0.1397 = 0.2853$$

The complete results for the overall positive-ideal and negative-ideal distances of each alternative are presented in Table 7.24.

TABLE 7.24

Overall Positive-Ideal and Negative-Ideal Distances for PROBID Walk-Through.

Alternatives	$S_{i(pos-ideal)}$	$S_{i(neg-ideal)}$
A1	0.2853	0.3386
A2	0.3150	0.3867
A3	0.3809	0.1593
A4	0.3925	0.3719
A5	0.3171	0.2728

Step 6. Compute the performance score (P_i) of each alternative and the alternative with the largest P_i is top-ranked and recommended to the decision-maker.

$$R_i = \frac{S_{i(pos-ideal)}}{S_{i(neg-ideal)}}$$

$$P_i = \frac{1}{1+R_i^2} + S_{i(avg)} \tag{7.59}$$

NUMERICAL CALCULATIONS

For instance, for $i = 1$:

$$R_1 = \frac{0.2853}{0.3386} = 0.8425$$

$$P_1 = \frac{1}{1 + 0.8425^2} + 0.1084 = 0.6933$$

Similarly, the performance scores for all alternatives are calculated as $P_1 = 0.6933$, $P_2 = 0.7401$, $P_3 = 0.2431$, $P_4 = 0.6377$ and $P_5 = 0.5244$. Accordingly, the ranking is A2 > A1 > A4 > A5 > A3, with A2 being the top-ranked alternative by PROBID.

The advantages of PROBID are as follows: (1) It uses several tiers of PIS and NIS, mitigating the impact of outliers, as no single outlier can dominate the references (Park et al., 2023). (2) It balances out extremes by factoring in the centroid (average) of all alternatives. (3) Rank reversal is mitigated due to the use of multiple reference points, which can provide a more comprehensive evaluation framework.

The limitations of PROBID are as follows: (1) Computing distances to multiple tiers of ideal solutions and an average solution is computationally intensive compared to other reference-type methods, making it potentially less efficient when handling a large ACM. (2) It is not completely outlier-proof, as extreme outliers can still affect the formation of different ideal tiers and/or distort the average solution.

7.11 Summary

Main points in this chapter of reference-type MCDM methods are as follows:

1. This chapter describes and illustrates nine reference-type MCDM methods, each with distinct approaches to defining reference solutions and computing performance scores. Listed chronologically, these methods are TOPSIS, GRA, VIKOR, EDAS, MABAC, CODAS, PIV, MARCOS and PROBID.
2. The methods covered are chosen based on either their widespread adoption (e.g., TOPSIS and VIKOR) or their recent development (e.g., MARCOS and PROBID).

3. Reference-type MCDM methods rank alternatives by comparing them to specific reference points (or solutions such as PIS and NIS) derived from the ACM. These reference points serve as benchmarks for evaluating the performance of alternatives.
4. Each method is clearly presented with its theoretical foundations, step-by-step algorithm and numerical calculations to enhance reader comprehension. The calculations are performed using a common ACM dataset, allowing for easy comparison across different methods.
5. Each method has its own advantages and limitations, as presented in its respective section.
6. Rank reversal is a common limitation, as adding or removing alternatives can modify the reference solutions. As a result, the relative rankings of existing alternatives may shift based on the updated reference points.
7. Some methods primarily rely on distances to PIS and/or NIS (e.g., TOPSIS), while some are based on distances to an arithmetic or geometric mean solution (e.g., MABAC). Additionally, some methods incorporate both ideal and mean solutions (e.g., PROBID).
8. Understanding the principles of each method can help decision-makers choose the most appropriate one based on the specific decision-making context.
9. Mastering these reference-type MCDM methods allows decision-makers, researchers and practitioners to effectively apply them to engineering and scientific applications that require systematic decision analysis.
10. Finally, as anticipated, the ranking results from the nine MCDM methods applied to the illustrated example show variation in the top-ranked alternative. While some methods consistently prioritize the same alternative, others yield different rankings due to their distinct principles and ranking mechanisms. For the illustrated example, A2 is the top-ranked alternative according to TOPSIS, CODAS, MARCOS and PROBID; A4 is ranked highest by GRA, EDAS, MABAC and PIV; and A1 is the top-ranked alternative under VIKOR. Generally, while A2 and A4 appear frequently at the top, the ranking order of other alternatives varies across methods, as seen in each individual section. These ranking results cannot be generalized and it is better to try several MCDM methods for any application.

References

Baydaş, M., Kavacık, M., & Wang, Z. (2024). Interpreting the determinants of sensitivity in MCDM methods with a new perspective: An application on e-scooter

selection with the PROBID method. *Spectrum of Engineering and Management Sciences, 2*(1), 17–35.
Behzadian, M., Otaghsara, S. K., Yazdani, M., & Ignatius, J. (2012). A state-of the-art survey of TOPSIS applications. *Expert Systems with Applications, 39*(17), 13051–13069.
Chen, L., & Gou, X. (2022). The application of probabilistic linguistic CODAS method based on new score function in multi-criteria decision-making. *Computational and Applied Mathematics, 41*, 1–25.
Demir, G., Chatterjee, P., Kadry, S., Abdelhadi, A., & Pamučar, D. (2024). Measurement of Alternatives and Ranking According to Compromise Solution (MARCOS) method: A comprehensive bibliometric analysis. *Decision Making: Applications in Management and Engineering, 7*(2), 313–336.
Deng, J.-L. (1982). Control problems of grey systems. *Systems & Control Letters, 1*, 288–294.
Gul, M., Celik, E., Aydin, N., Gumus, A. T., & Guneri, A. F. (2016). A state of the art literature review of VIKOR and its fuzzy extensions on applications. *Applied Soft Computing, 46*, 60–89.
Hwang, C.-L., & Yoon, K. (1981). *Multiple attribute decision making: Methods and applications*. Berlin, Germany: Springer.
Keshavarz Ghorabaee, M., Zavadskas, E. K., Olfat, L., & Turskis, Z. (2015). Multi-criteria inventory classification using a new method of evaluation based on distance from average solution (EDAS). *Informatica, 26*(3), 435–451.
Keshavarz Ghorabaee, M., Zavadskas, E. K., Turskis, Z., & Antucheviciene, J. (2016). A new combinative distance-based assessment (CODAS) method for multi-criteria decision-making. *Economic Computation & Economic Cybernetics Studies & Research, 50*(3).
Kizielewicz, B., Shekhovtsov, A., & Sałabun, W. (2023). pymcdm – the universal library for solving multi-criteria decision-making problems. *SoftwareX, 22*, 101368.
Kundakçı, N. (2016). Personnel selection with grey relational analysis. *Management Science Letters, 6*(5), 351–360.
Liao, H., & Xu, Z. (2013). A VIKOR-based method for hesitant fuzzy multi-criteria decision making. *Fuzzy Optimization and Decision Making, 12*(4), 373–392.
Martinez-Morales, J. D., Pineda-Rico, U., & Stevens-Navarro, E. (2010). Performance comparison between MADM algorithms for vertical handoff in 4G networks. In *2010 7th International Conference on Electrical Engineering Computing Science and Automatic Control*. IEEE.
Mufazzal, S., & Muzakkir, S. M. (2018). A new Multi-Criterion Decision Making (MCDM) method based on proximity indexed value for minimizing rank reversals. *Computers & Industrial Engineering, 119*, 427–438.
Nabavi, S. R., Guo, Z., & Wang, Z. (2025). Machine learning-assisted surrogate modeling with multi-objective optimization and decision-making of a steam methane reforming reactor. *arXiv preprint arXiv:2507.07641*. https://doi.org/10.48550/arXiv.2507.07641
Nabavi, S. R., Jafari, M. J., & Wang, Z. (2023). Deep learning aided multi-objective optimization and multi-criteria decision making in thermal cracking process for olefines production. *Journal of the Taiwan Institute of Chemical Engineers, 152*, 105179.

Nabavi, S. R., Wang, Z., & Rodríguez, M. L. (2024). Multi-objective optimization and multi-criteria decision-making approach to design a multi-tubular packed-bed membrane reactor in oxidative dehydrogenation of ethane. *Energy & Fuels, 39*(1), 491–503.

Opricovic, S. (1998). Multicriteria optimization of civil engineering systems. *Faculty of Civil Engineering*, 5–21.

Opricovic, S., & Tzeng, G. H. (2004). Compromise solution by MCDM methods: A comparative analysis of VIKOR and TOPSIS. *European Journal of Operational Research, 156*(2), 445–455.

Özçelik, F., & Öztürk, B. A. (2014). Evaluation of banks' sustainability performance in Turkey with grey relational analysis. *The Journal of Accounting and Finance, 63*, 189–209.

Pamučar, D., & Ćirović, G. (2015). The selection of transport and handling resources in logistics centers using Multi-Attributive Border Approximation Area Comparison (MABAC). *Expert Systems with Applications, 42*(6), 3016–3028.

Park, M., Wang, Z., Li, L., & Wang, X. (2023). Multi-objective building energy system optimization considering EV infrastructure. *Applied Energy, 332*, 120504.

Sałabun, W., Wątróbski, J., & Shekhovtsov, A. (2020). Are MCDA methods benchmarkable? A comparative study of TOPSIS, VIKOR, COPRAS and PROMETHEE ii methods. *Symmetry, 12*(9), 1549.

Scarlat, E. (2015). The hybrid intelligent systems design using grey systems theory. In *2015 IEEE International Conference on Grey Systems and Intelligent Services (GSIS)* (pp. 19–28). IEEE.

Shekhovtsov, A., & Sałabun, W. (2020). A comparative case study of the VIKOR and TOPSIS rankings similarity. *Procedia Computer Science, 176*, 3730–3740.

Song, Q., & Jamalipour, A. (2005). Network selection in an integrated wireless LAN and UMTS environment using mathematical modeling and computing techniques. *IEEE Wireless Communications, 12*(3), 42–48.

Stević, Ž., Pamučar, D., Puška, A., & Chatterjee, P. (2020). Sustainable supplier selection in healthcare industries using a new MCDM method: Measurement of Alternatives and Ranking According to Compromise Solution (MARCOS). *Computers & Industrial Engineering, 140*, 106231.

Wang, Z., Li, J., Rangaiah, G. P., & Wu, Z. (2022). Machine learning aided multi-objective optimization and multi-criteria decision making: Framework and two applications in chemical engineering. *Computers & Chemical Engineering, 165*, 107945.

Wang, Z., Nabavi, S. R., & Rangaiah, G. P. (2023). Selected multi-criteria decision-making methods and their applications to product and system design. In *Optimization methods for product and system design* (pp. 107–138). Singapore: Springer Nature Singapore.

Wang, Z., Nabavi, S. R., & Rangaiah, G. P. (2024). Multi-criteria decision making in chemical and process engineering: Methods, progress and potential. *Processes, 12*(11), 2532.

Wang, Z., Rangaiah, G. P., & Wang, X. (2021). Preference ranking on the basis of ideal-average distance method for multi-criteria decision-making. *Industrial & Engineering Chemistry Research, 60*(30), 11216–11230.

Zhang, S. F., Liu, S. Y., & Zhai, R. H. (2011). An extended GRA method for MCDM with interval-valued triangular fuzzy assessments and unknown weights. *Computers & Industrial Engineering, 61*(4), 1336–1341.

Exercises

E7.1 You are given an ACM consisting of four alternatives (A1, A2, A3 and A4) and three criteria (C1, C2 and C3), with the following values:

Alternatives	C1	C2	C3
A1	0.93	600	8.25
A2	0.51	700	6.33
A3	0.77	500	3.16
A4	0.82	400	2.98

Criteria C1 and C2 are the benefit criteria to be maximized, whereas C3 is a cost criterion to be minimized. The weights for the criteria are $w_1 = 0.5$, $w_2 = 0.3$ and $w_3 = 0.2$. Apply all (or some of) the reference-type MCDM methods covered in this chapter to rank the alternatives. Identify the top-ranked alternative by each method and compare them.

E7.2 You are given an ACM consisting of five alternatives (A1, A2, A3, A4 and A5) and four criteria (C1, C2, C3 and C4), with the following values:

Alternatives	C1	C2	C3	C4
A1	234	0.122	90.3	0.069
A2	179	0.641	13.2	0.032
A3	398	0.782	67.1	0.191
A4	273	0.979	49.8	0.264
A5	278	0.543	86.8	0.219

Criteria C2 and C3 are the benefit criteria to be maximized, whereas C1 and C4 are the cost criteria to be minimized. The weights for the criteria are $w_1 = 0.3$, $w_2 = 0.4$, $w_3 = 0.2$ and $w_4 = 0.1$. Apply all (or some of) the reference-type MCDM methods covered in this chapter to rank the alternatives. Identify the top-ranked alternative by each method and compare them. *Note that calculations in this and subsequent examples increase with the increasing number of alternatives and criteria. Hence, readers are advised to implement the calculations of each method in a Microsoft Excel worksheet.*

E7.3 You are given an ACM consisting of seven alternatives (A1, A2, A3, A4, A5, A6 and A7) and five criteria (C1, C2, C3, C4 and C5), with the following values:

Alternatives	C1	C2	C3	C4	C5
A1	3405	87.4	0.245	0.105	4.2
A2	2159	45.2	0.521	0.187	3.7
A3	4782	72.1	0.684	0.274	5.1
A4	3594	33.9	0.319	0.143	2.9
A5	2911	94.3	0.753	0.238	4.8
A6	4100	59.7	0.602	0.194	3.4
A7	3317	80.6	0.438	0.165	4.7

Criteria C1, C2 and C3 are the benefit criteria to be maximized, whereas C4 and C5 are the cost criteria to be minimized. The weights for the criteria are $w_1 = 0.25$, $w_2 = 0.2$, $w_3 = 0.3$, $w_4 = 0.15$ and $w_5 = 0.1$. Apply all (or some of) the reference-type MCDM methods covered in this chapter to rank the alternatives. Identify the top-ranked alternative by each method and compare them.

E7.4 You are given an ACM consisting of ten alternatives (A1, A2, A3, A4, A5, A6, A7, A8, A9 and A10) and six criteria (C1, C2, C3, C4, C5 and C6), with the following values:

Alternatives	C1	C2	C3	C4	C5	C6
A1	575	0.125	63.8	0.0215	12.3	3.5
A2	432	0.315	89.2	0.0382	15.7	2.9
A3	689	0.498	74.5	0.0497	18.2	4.1
A4	540	0.276	95.3	0.0328	14.9	3.8
A5	478	0.605	70.4	0.0409	16.5	4.5
A6	615	0.451	82.6	0.0462	17.9	3.7
A7	503	0.333	88.9	0.0375	13.8	3.2
A8	389	0.254	59.1	0.0293	10.5	2.8
A9	455	0.394	76.3	0.0319	14.2	4.3
A10	612	0.512	81.9	0.0415	15.1	3.6

Criteria C1 and C4 are the benefit criteria to be maximized, whereas C2, C3, C5 and C6 are the cost criteria to be minimized. The weights for the criteria are $w_1 = 0.2$, $w_2 = 0.15$, $w_3 = 0.25$, $w_4 = 0.15$, $w_5 = 0.15$ and $w_6 = 0.1$. Apply all (or some of) the reference-type MCDM methods covered in this chapter to rank the alternatives. Identify the top-ranked alternative by each method and compare them.

E7.5 Describe the similarities and differences between TOPSIS and PIV.

E7.6 Describe the similarities and differences between EDAS and CODAS.

E7.7 Describe the similarities and differences between MABAC and PROBID.

8

Multi-Criteria Decision-Making – Aggregation-Type Methods

8.1 Overview

For a given alternatives-criteria matrix (ACM), multi-criteria decision-making (MCDM) serves as an effective approach for selecting the best alternative(s) from the matrix (Baydaş et al., 2024). Among the diverse MCDM methodologies, this chapter focuses on aggregation-type MCDM methods. These methods compute an overall performance score for each alternative by aggregating the values of all criteria using different principles or formulations (Wang et al., 2024). The aggregation process typically follows either an additive, multiplicative, or hybrid approach, ensuring that the final ranking of alternatives reflects the collective impact of all criteria.

This chapter covers eight aggregation-type MCDM methods, detailing their principles and algorithms. These methods are selected based on either their widespread popularity for various applications or their recent development. Each method is elucidated through a simple numerical example, ensuring that readers gain a clear understanding of its inner workings.

The following sections in this chapter present the eight aggregation-type MCDM methods, in chronological order. Readers need not study them sequentially and can read any of them (independent of others).

- Section 8.2 Simple Additive Weighting (SAW)
- Section 8.3 Multiplicative Exponent Weighting (MEW)
- Section 8.4 Analytic Hierarchy Process (AHP)
- Section 8.5 Analytic Network Process (ANP)
- Section 8.6 Complex Proportional Assessment (COPRAS)
- Section 8.7 Multi-Objective Optimization on the Basis of Ratio Analysis (MOORA)
- Section 8.8 Faire Un Choix Adéquat (FUCA)

DOI: 10.1201/9781003530435-8

Section 8.9 Weighted Aggregated Sum Product Assessment (WASPAS)
Section 8.10 Summary

Our Microsoft Excel-based program, EMCDM, described in Chapter 10, includes SAW, COPRAS, MOORA and FUCA methods.

The learning outcomes of this chapter are to (1) describe the principles and algorithms of aggregation-type MCDM methods covered, (2) apply any of these methods to ACM of an application and (3) select one or more aggregation-type MCDM methods for solving MCDM problems.

Unless otherwise stated, the notation used throughout this chapter is as follows:

- $i \in \{1,2,\ldots,m\}$ and $j \in \{1,2,\ldots,n\}$, where m is the number of rows (i.e., alternatives) and n is the number of columns (i.e., criteria) in the ACM.
- f_{ij} is the value of the jth criterion for the ith alternative in the original ACM.
- F_{ij} is the normalized value of f_{ij}.
- w_j is the weight assigned to the jth criterion and $\sum_{j=1}^{n} w_j = 1$.
- v_{ij} is the value of the jth criterion for the ith alternative in the weighted normalized ACM.

Additionally, the ACM presented in Table 8.1 is employed to walk through the steps of the aggregation-type MCDM methods covered in this chapter. This ACM is made up of four alternatives (i.e., $m = 4$ and $i \in \{1,2,3,4\}$) and three criteria (i.e., $n = 3$ and $j \in \{1,2,3\}$). Out of the three criteria (C1, C2 and C3), C1 and C2 are benefit criteria to be maximized, while C3 is a cost criterion to be minimized. For illustration, the three criteria are assigned weights as follows: $w_1 = 0.25$, $w_2 = 0.33$ and $w_3 = 0.42$. All four alternatives (A1, A2, A3 and A4) are non-dominated, which means that the improvement in one

TABLE 8.1

ACM Employed for Illustration of Aggregation-Type MCDM Methods (C1 and C2 for Maximization; C3 for Minimization).

Alternatives	C1 (Max)	C2 (Max)	C3 (Min)
A1	0.93	600	8.25
A2	0.51	700	6.33
A3	0.77	500	3.16
A4	0.82	400	2.98
Weight	0.25	0.33	0.42

criterion will inevitably result in degradation in at least one other criterion (Wang et al., 2023).

8.2 Simple Additive Weighting (SAW)

The SAW method, also known as the WSM, described in MacCrimmon (1968), is one of the most fundamental and simple methods in MCDM. It is based on the principle that the top-ranked alternative is the one with the highest WS of normalized criteria values (Nabavi et al., 2023), after converting minimization or cost criteria into maximization or benefit type. The method assumes that the criteria are compensatory, which means that a lower score in one criterion can be offset by a higher score in another. The steps in SAW are as follows:

Step 1. Normalize the original ACM with m rows (i.e., alternatives) and n columns (i.e., criteria) using the max normalization method. For a maximization criterion, the normalized value, F_{ij}, is obtained as follows:

$$F_{ij} = \frac{f_{ij}}{\max_{k \in \{1,2,\ldots,m\}} f_{kj}} \tag{8.1}$$

For a minimization criterion, the normalized value, F_{ij}, is obtained as follows:

$$F_{ij} = \frac{\min_{k \in \{1,2,\ldots,m\}} f_{kj}}{f_{ij}} \tag{8.2}$$

The above normalization ensures that all minimization criteria are converted to the maximization type, indicating that a higher F_{ij} always indicates a better performance (Nabavi et al., 2024).

NUMERICAL CALCULATIONS (USING ACM FROM TABLE 8.1)

For instance, for $i = 1$ and $j = 2$:

$$F_{12} = \frac{600}{700} = 0.8571$$

Similar calculations are performed for all other values in the ACM across the four alternatives and three criteria. The complete results, forming the normalized ACM, are shown in Table 8.2.

TABLE 8.2

Normalized ACM for SAW Walk-Through.

Alternatives	C1	C2	C3
A1	1	0.8571	0.3612
A2	0.5484	1	0.4708
A3	0.8280	0.7143	0.9430
A4	0.8817	0.5714	1

Step 2. Construct the weighted normalized ACM by multiplying each normalized criterion value by its corresponding weight w_j.

$$v_{ij} = F_{ij} \times w_j \tag{8.3}$$

NUMERICAL CALCULATIONS

For instance, for $i = 1$ and $j = 2$:

$$v_{12} = F_{12} \times w_2 = 0.8571 \times 0.33 = 0.2829$$

Likewise, calculations are carried out for all other values in the normalized ACM, using their respective assigned weights. The results, which constitute the weighted normalized ACM, are displayed in Table 8.3.

TABLE 8.3

Weighted Normalized ACM for SAW Walk-Through.

Alternatives	C1	C2	C3
A1	0.2500	0.2829	0.1517
A2	0.1371	0.3300	0.1977
A3	0.2070	0.2357	0.3961
A4	0.2204	0.1886	0.4200

Step 3. Compute the performance score (P_i) for each alternative, which is obtained by summing its weighted normalized values.

$$P_i = \sum_{j=1}^{n} v_{ij} \tag{8.4}$$

Finally, the alternatives are ranked in descending order of P_i, with the highest value indicating the top-ranked alternative recommended to the decision-maker.

NUMERICAL CALCULATIONS

For instance, for $i = 1$:

$$P_1 = \sum_{j=1}^{3} v_{1j} = 0.2500 + 0.2829 + 0.1517 = 0.6846$$

Similarly, the performance scores of all alternatives are calculated as $P_1 = 0.6846$, $P_2 = 0.6648$, $P_3 = 0.8388$ and $P_4 = 0.8290$. Accordingly, the ranking is A3 > A4 > A1 > A2, with A3 being the top-ranked alternative by SAW.

The advantages of SAW are as follows: (1) It is one of the simplest MCDM methods, requiring only the sum of the normalized weighted criteria values, which makes it very straightforward to understand and implement (Wang et al., 2022). (2) Its computational procedure is relatively transparent, allowing decision-makers to easily observe how each criterion and weight contributes to the overall score. (3) Due to its simplicity, it can be readily applied in many fields, such as engineering, project selection and resource allocation, where a quick ranking of alternatives is needed.

The limitations of SAW are as follows: (1) Similar to other MCDM methods, rank reversal can occur if new alternatives are added or existing ones are removed, potentially affecting the stability of the ranking results. (2) It treats all performance values additively; high performance on one criterion is possible to completely compensate for low performance on another, which might be undesirable in certain decision-making scenarios.

8.3 Multiplicative Exponent Weighting (MEW)

The MEW method, also known as the weighted product method (WPM), was introduced in Miller and Starr (1969). It is an MCDM technique that evaluates alternatives based on multiplicative aggregation. Unlike the SAW method, the MEW method uses the product of normalized criteria values raised to the power of their respective weights. The alternative that has a larger product value is ranked higher.

Step 1. Normalize the original ACM with m rows (i.e., alternatives) and n columns (i.e., criteria) using the max normalization method. This step is the same as Step 1 of the SAW method. For the sake of brevity, the normalization equations and numerical calculations (same as Table 8.2) are not repeated here.

Step 2. Calculate the performance score (P_i) for each alternative using the multiplicative aggregation equation as follows:

$$P_i = \prod_{j=1}^{n} F_{ij}^{w_j} \tag{8.5}$$

The alternative with the largest P_i is top-ranked and recommended to the decision-maker.

NUMERICAL CALCULATIONS

For instance, for $i = 1$ and using normalized ACM in Table 8.2:

$$P_1 = \prod_{j=1}^{n} F_{1j}^{w_j} = 1.0000^{0.25} \times 0.8571^{0.33} \times 0.3612^{0.42} = 0.6197$$

Similarly, the performance scores for all alternatives are calculated as $P_1 = 0.6197$, $P_2 = 0.6271$, $P_3 = 0.8329$ and $P_4 = 0.8056$. Accordingly, the ranking is A3 > A4 > A2 > A1, with A3 being the top-ranked alternative by MEW. In this example, the top-ranked alternative, namely A3, is the same for both SAW and MEW. However, this may not always be the case in other applications.

The advantages of MEW are as follows: (1) As it uses a multiplicative aggregation, an alternative must perform reasonably well in each criterion to achieve a high overall performance score; this property helps avoid the possible excessive compensation (i.e., very high performance in one criterion fully offsets poor performance in another, as can occur in SAW). (2) Conceptually, MEW is straightforward and decision-makers can easily understand the mechanics of how each criterion contributes to the final performance score. (3) Due to its multiplicative structure, MEW is sensitive to extreme degradation in a particular criterion, as a low value can substantially reduce the overall product. This characteristic makes it well-suited for risk-averse applications.

The limitations of MEW are as follows: (1) If the performance on a single criterion is zero (or extremely small) for a particular alternative, the product becomes zero (or very close to zero), effectively disqualifying that alternative regardless of its performance on other criteria. (2) Like other MCDM methods, rank reversal can still occur if new alternatives are introduced or existing ones are deleted, since the normalization step and multiplicative aggregation can shift the overall scale and relative scores among alternatives.

8.4 Analytic Hierarchy Process (AHP)

The AHP method, originally developed by Saaty (1977, 1990), is based on decision-makers' subjective pairwise comparisons. To facilitate this, Saaty's fundamental 1–9 scale (Table 8.4) is typically employed. The AHP method begins by constructing a decision hierarchy, organizing the overall goal, n criteria and m alternatives into a structured model, as illustrated in Figure 8.1. Decision-makers then perform subjective pairwise comparisons to assess the relative importance of each pair of criteria using Saaty's 1–9 scale. These comparisons are compiled into an $n \times n$ pairwise comparison matrix. Following this, the eigenvalue method is used to find the priority eigenvector from the matrix, which then determines the criteria weights (Saaty, 1990). Instead of the eigenvalue method, there is a simpler approximation technique (based on sum normalization of the pairwise comparison matrix) to calculate the criteria weights, which is adopted in many applications (e.g., Teknomo, 2006; Karim and Karmaker, 2016; Wang et al., 2020). The procedure for determining criteria weights follows the same steps outlined in the AHP section of Chapter 6. If the criteria weights have already been explicitly predefined by decision-makers, this step may be omitted. Note that the determination of criteria weights by the AHP method does not require ACM.

In addition to determining the criteria weights, it is crucial to recognize that the AHP method can be directly utilized for decision-making (i.e., act as an MCDM method). For this, the conventional procedure is that decision-makers conduct subjective pairwise comparisons among alternatives with respect to each criterion, leading to n number of $m \times m$ pairwise comparison matrices. According to the original AHP method (Saaty, 1990), these pairwise comparisons among alternatives are purely subjective, sourced directly from the judgments of the decision-makers. For each $m \times m$

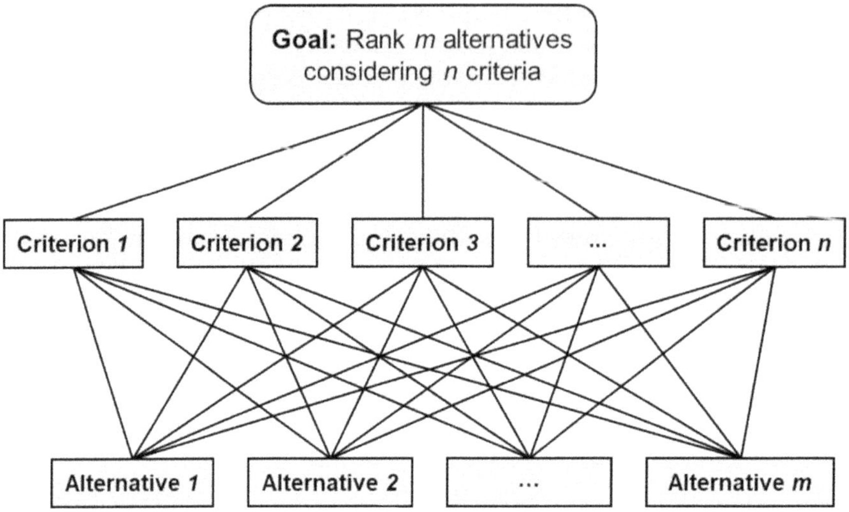

FIGURE 8.1
Illustration of the AHP decision hierarchy of goal, criteria and alternatives in top, middle and bottom levels.

TABLE 8.4
Saaty's 1–9 Pairwise Comparison Scale.

Scale	Compare Between Events x and y
1	Equal importance of x against y
3	Moderate importance of x against y
5	Essential or strong importance of x against y
7	Very strong importance of x against y
9	Extreme importance of x against y
2, 4, 6, 8	Intermediate value between adjacent scales
Any decimal within [1, 9], for example, 2.3 and 5.1	Finer scale
Reciprocals, for example, 1/3 and 1/7	If event x has one of the above numbers assigned to it when compared with event y, then y has the reciprocal value when compared with x.

pairwise comparison matrix, AHP calculates the local priority vector (sized $m \times 1$) for alternatives with respect to the corresponding criterion using the eigenvalue method, analogous to the calculation of the criteria weights at the beginning.

Notably, a typical ACM containing objective quantitative data on different measurement scales (e.g., Table 8.1), rather than the subjective pairwise comparisons using Saaty's 1–9 scale, is neither required nor immediately compatible with the AHP acting as an MCDM method. However, mapping the objective quantitative data (e.g., Table 8.1) to Saaty's 1–9 scale and constructing n number of $m \times m$ pairwise comparison matrices is feasible (Si et al., 2016).

Subsequently, AHP iterates through all n criteria, calculates and amalgamates these local priority vectors ($m \times 1$) into a local priority matrix ($m \times n$), in which each row indicates the priority of an alternative with respect to all n criteria. This $m \times n$ local priority matrix is then multiplied by the $n \times 1$ criteria weights vector (previously calculated by AHP or predefined by decision-makers), resulting in an $m \times 1$ vector representing the global priorities of all alternatives. The alternative with the highest global priority value is top-ranked and recommended to decision-makers.

The main procedural steps of the AHP method are summarized below. For completeness, we will go through the full AHP steps along with numerical calculations, starting with determining the criteria weights, followed by calculating the local priority vectors of alternatives with respect to each criterion and finally aggregating the results to rank the alternatives. In the following, Steps 1–3 are for the criteria weights and Steps 4 and 5 are for ranking the alternatives.

Step 1. Construct the $n \times n$ pairwise comparison matrix for n criteria as follows:

$$\mathbf{M}_{\mathbf{A}} = \begin{bmatrix} a_{11} & a_{12} & a_{13} & \cdots & a_{1n} \\ a_{21} & a_{22} & a_{23} & \cdots & a_{2n} \\ a_{31} & a_{32} & a_{33} & \cdots & a_{3n} \\ \vdots & \vdots & \vdots & \ddots & \vdots \\ a_{n1} & a_{n2} & a_{n3} & \cdots & a_{nn} \end{bmatrix} \tag{8.6}$$

- Here, $i, j \in \{1, 2, \ldots, n\}$, each entry a_{ij} represents the relative importance of criterion i compared to criterion j using Saaty's 1–9 scale (in Table 8.4).
- The leading diagonal entries ($a_{11}, a_{22}, a_{33}, \ldots, a_{nn}$) of this matrix are always equal to unity, that is, the relative importance of a criterion compared to itself is always 1.
- The following equation always holds true (i.e., entries below the leading diagonal are the reciprocals of the corres ponding entries above it), as per the last row of Table 8.4.

$$a_{ij} = \frac{1}{a_{ji}} \tag{8.7}$$

- For instance, if we consider that the relative importance of criterion 1 versus criterion 2 is moderate (i.e., $a_{12} = 3$), then the relative importance of criterion 2 to criterion 1 is its reciprocal (i.e., $a_{21} = 1/3$). Thus, only $\frac{n(n-1)}{2}$ comparisons are needed to populate half of the matrix, with values of the other half obtained through simple reciprocals.

NUMERICAL CALCULATIONS

Instead of using the predefined weights presented in the last row of Table 8.1, suppose that the decision-maker provides the following subjective assessments for the three criteria (C1, C2 and C3) using Saaty's 1–9 scale:

- C1 is moderately more important than C2 (i.e., $a_{12} = 3$ and $a_{21} = 1/3$), meaning C1 is three times as important as C2, while C2 holds only 1/3 the importance of C1.
- C1 is strongly more important than C3 (i.e., $a_{13} = 5$ and $a_{31} = 1/5$).
- C2 is rated between moderate and strong in importance relative to C3 (i.e., $a_{23} = 4$ and $a_{32} = 1/4$).

Based on these assessments, the 3×3 pairwise comparison matrix is as follows:

$$\mathbf{M}_{\mathrm{A}} = \begin{bmatrix} a_{11} & a_{12} & a_{13} \\ a_{21} & a_{22} & a_{23} \\ a_{31} & a_{32} & a_{33} \end{bmatrix} = \begin{bmatrix} 1 & 3 & 5 \\ 1/3 & 1 & 4 \\ 1/5 & 1/4 & 1 \end{bmatrix}$$

Step 2. Compute the priority vector (i.e., criteria weights in this step), using the eigenvalue method by solving the following equation.

$$\mathbf{M}_{\mathrm{A}}\, \mathbf{w}^{*} = \lambda_{\max}\, \mathbf{w}^{*} \tag{8.8}$$

Here, $\lambda_{\max}$ is the largest eigenvalue (known as the principal eigenvalue) of the pairwise comparison matrix and $\mathbf{w}^{*}$ is the corresponding principal eigenvector. The entries of this principal eigenvector are then normalized to sum to 1, yielding the criteria weights in this step, as follows:

$$w_j = \frac{w_j^{*}}{\sum_{k=1}^{n} w_k^{*}}, \quad j = 1, 2, \ldots, n \tag{8.9}$$

Here, w_j^* and w_k^* represent the jth and k^{th} entries of the principal eigenvector $\mathbf{w}^*$, respectively, while w_j denotes the final weight of the criterion j.

NUMERICAL CALCULATIONS

Solve the following equation using computational tools such as WolframAlpha (https://www.wolframalpha.com/input?i=eigenvalue+calculator) or Python code:

$$\begin{bmatrix} 1 & 3 & 5 \\ 1/3 & 1 & 4 \\ 1/5 & 1/4 & 1 \end{bmatrix} \mathbf{w}^* = \lambda_{\max} \mathbf{w}^*$$

Use of WolframAlpha gives the principal eigenvalue $\lambda_{\max} = 3.086$ and the corresponding principal eigenvector $\mathbf{w}^* = \begin{bmatrix} 6.6943 \\ 2.9876 \\ 1 \end{bmatrix}$. Note that when using different computational tools, the computed principal eigenvector may be scaled by a constant factor.

Next, apply sum normalization to the principal eigenvector to derive the weights of the three criteria (C1, C2 and C3), as follows:

$$w_1 = \frac{w_1^*}{\sum_{k=1}^{3} w_k^*} = \frac{6.6943}{6.6943 + 2.9876 + 1} = 0.6267$$

$$w_2 = \frac{w_2^*}{\sum_{k=1}^{3} w_k^*} = \frac{2.9876}{6.6943 + 2.9876 + 1} = 0.2797$$

$$w_3 = \frac{w_3^*}{\sum_{k=1}^{3} w_k^*} = \frac{1}{6.6943 + 2.9876 + 1} = 0.0936$$

As expected, these weights very much depend on the values given by decision-makers for pairwise comparison of criteria. Further, the above weights are different from the arbitrarily assumed weight values in Table 8.1.

TABLE 8.5

Saaty's RI Values. For $n > 10$, See Podvezko (2009).

n	1	2	3	4	5	6	7	8	9	10
RI	0	0	0.58	0.90	1.12	1.24	1.32	1.41	1.45	1.49

Step 3. Check the consistency of the pairwise comparison matrix of n criteria by calculating its CR.

$$CI = \frac{\lambda_{max} - n}{n-1}, CR = \frac{CI}{RI} \tag{8.10}$$

Here, CI is the consistency index and RI (shown in Table 8.5) is the random CI value provided by Saaty (1990), who computed it by averaging the CI values from numerous randomly generated $n \times n$ reciprocal matrices using Saaty's 1–9 scale. If $CR \leq 0.1$, consistency is considered acceptable. If not, decision-makers should review and revise their given values for pairwise comparison of criteria.

In addition, for a perfectly consistent pairwise comparison matrix (where each comparison a_{ij} between criterion i and criterion j exactly equals the ratio of their resulting weights w_i / w_j, which also implies that the transitive property holds, i.e., $a_{ij}a_{jk} = a_{ik}$ for all criteria i, j and k), Saaty (1990) proved that the largest eigenvalue of this matrix is equal to the number of criteria, that is, $\lambda_{\max} = n$.

Note on Consistency Analysis: Since we are dealing with human judgment for a pairwise comparison matrix, it is inherently subjective and may sometimes be vague or inconsistent. For example, if a decision-maker prefers criterion 1 over criterion 2 and criterion 2 over criterion 3, then logically, he or she should prefer criterion 1 over criterion 3. This logical principle is known as the transitive property. However, in practice, due to cognitive biases or judgmental inconsistencies, decision-makers may unknowingly violate this property to some extent.

NUMERICAL CALCULATIONS

In this example, with $n = 3$, refer to Table 8.5 and find the corresponding $RI = 0.58$.

$$CI = \frac{\lambda_{\max} - n}{n-1} = \frac{3.086 - 3}{3-1} = 0.0429$$

$$CR = \frac{CI}{RI} = \frac{0.0429}{0.58} = 0.074$$

Since the calculated CR satisfies $CR \leq 0.1$, the decision-maker's subjective assessments for the three criteria are considered consistent and acceptable.

After obtaining the criteria weights, the following steps are for ranking the alternatives by the AHP method.

Step 4. Determine local priority scores of alternatives under each criterion, denoted as v_{ij} with $i \in \{1, 2, ..., m\}$ and $j \in \{1, 2, ..., n\}$.

With respect to each of the n criteria, a separate pairwise comparison matrix (size $m \times m$) is constructed, comparing all pairs of m alternatives, which essentially follows the same procedure as described in Steps 1–3 including consistency analysis. For example, the decision-maker assesses alternative 1 against alternative 2 under the first criterion and assigns a relative importance value using Saaty's 1–9 scale. This process is repeated for all pairs of m alternatives under each criterion. Once the n number of $m \times m$ pairwise comparison matrices is established, the local priority vectors (representing the relative priorities of alternatives for each criterion) are computed using the eigenvalue method as in Step 2.

At this stage, if the decision-maker does not have an ACM (such as Table 8.1) based on objective measurement data, they can continue to rely entirely on their subjective judgments, as in Step 1, to construct all pairwise comparison matrices for alternatives under each criterion (e.g., A1 is moderately more important/better than A4 under C1; A3 is strongly more important/better than A2 under C3).

If an ACM based on objective measurement data is available, since it is not directly compatible with the conventional AHP process that relies on Saaty's 1–9 scale, the decision-maker needs to first map the objective data into Saaty's 1–9 scale (Si et al., 2016). This is done either through subjective assessment, leveraging the decision-maker's domain knowledge to inspect the ACM and assign pairwise comparison values of alternatives under each criterion using the 1–9 scale, or by applying mathematical mapping techniques. After this mapping, the decision-maker constructs the pairwise comparison matrices for alternatives for each criterion. One commonly used mathematical mapping technique is the logarithmic transformation, whose steps are outlined below.

Firstly, normalize the original ACM with m rows (i.e., alternatives) and n columns (i.e., criteria) using the max normalization method.

For a maximization criterion, the normalized value F_{ij} is obtained as follows:

$$F_{ij} = \frac{f_{ij}}{\max_{k \in \{1,2,\ldots,m\}} f_{kj}} \tag{8.11}$$

For a minimization criterion, the normalized value F_{ij} is obtained as follows:

$$F_{ij} = \frac{\min_{k \in \{1,2,\ldots,m\}} f_{kj}}{f_{ij}} \tag{8.12}$$

The above normalization transforms all minimization criteria to the maximization type (i.e., a higher F_{ij} always indicates a better performance).

Next, for each criterion j, determine the largest ratio ($r_{max,j}$) and smallest ratio ($r_{min,j}$) after comparing each alternative's normalized value with the minimum F_{ij} in that criterion:

$$r_{max,j} = \frac{\max_i F_{ij}}{\min_i F_{ij}},\ i \in \{1,2,\ldots,m\}\ and\ j \in \{1,2,\ldots,n\} \tag{8.13}$$

$$r_{min,j} = \frac{\min_i F_{ij}}{\min_i F_{ij}} = 1,\ i \in \{1,2,\ldots,m\}\ and\ j \in \{1,2,\ldots,n\} \tag{8.14}$$

Here, $r_{min,j}$ always equals 1 because it is the ratio of the minimum F_{ij} to itself.

Then, for each criterion j, the pairwise comparison between alternative i and k (i.e., a_{ik}) is mapped into Saaty's 1–9 scale using the logarithmic transformation as follows:

$$\text{When } F_{ij} \geq F_{kj},\quad a_{ik} = \frac{ln\left(\frac{F_{ij}}{F_{kj}}\right) - ln\left(r_{min,j}\right)}{ln\left(r_{max,j}\right) - ln\left(r_{min,j}\right)} \times (9-1) + 1,\ i, k \in \{1,2,\ldots,m\} \tag{8.15}$$

For cases of $F_{ij} < F_{kj}$ when comparing F_{ij} over F_{kj}, its pairwise comparison value a_{ik} is obtained through the simple reciprocal of the comparison of F_{kj} over F_{ij}, since $a_{ik} = \frac{1}{a_{ki}}$.

The reason logarithmic transformation is commonly used in this context is that Saaty's 1–9 scale is meant to convey a ratio of importance (1 means equal, 3 means moderate, 9 means extreme, etc.). A logarithmic transformation preserves these multiplicative (ratio) relationships. For example, comparing $F_{ij} = 0.2$ and $F_{kj} = 0.1$, their ratio $\frac{F_{ij}}{F_{kj}}$ is 2; thus, $ln\left(\frac{F_{ij}}{F_{kj}}\right) = 0.693$; similarly, comparing $F_{ij} = 0.9$ and $F_{kj} = 0.45$, their ratio $\frac{F_{ij}}{F_{kj}}$ is also 2 and $ln\left(\frac{F_{ij}}{F_{kj}}\right) = 0.693$. Since both comparisons yield the same logarithmic value, the pairwise comparison value a_{ik} remains the same in both cases.

Once the pairwise comparison matrices for all alternatives under each criterion are constructed, the eigenvalue method equation from Step 2 can be reused to compute the local priority scores (v_{ij}), which are derived by applying the sum normalization to the obtained principal eigenvector. In addition, the equation from Step 3 is reused to calculate the CRs of the matrices.

NUMERICAL CALCULATIONS (USING ACM FROM TABLE 8.1)

Apply the max normalization method; the original ACM is normalized and shown in Table 8.6.

TABLE 8.6

Normalized ACM for AHP Walk-Through.

Alternatives	C1	C2	C3
A1	1	0.8571	0.3612
A2	0.5484	1	0.4708
A3	0.8280	0.7143	0.9430
A4	0.8817	0.5714	1

For instance, for $j = 1$:

$$r_{max,1} = \frac{\max_i F_{i1}}{\min_i F_{i1}} = \frac{1}{0.5484} = 1.8235$$

$$r_{min,1} = \frac{\min_i F_{i1}}{\min_i F_{i1}} = 1$$

Next, under $j = 1$, let us compare A1's normalized value (F_{11}) over A3's normalized value (F_{31}) and perform a logarithmic transformation:

$$a_{13} = \frac{ln\left(\frac{F_{11}}{F_{31}}\right) - ln\left(r_{min,1}\right)}{ln\left(r_{max,1}\right) - ln\left(r_{min,1}\right)} \times (9-1)+1$$
$$= \frac{ln\left(\frac{1}{0.8280}\right) - ln(1)}{ln(1.8235) - ln(1)} \times (9-1)+1$$
$$= 3.5140$$

Symmetrically, we can obtain that $a_{31} = \frac{1}{a_{13}} = \frac{1}{3.5140} = 0.2846$.

Likewise, the complete pairwise comparisons for all four alternatives under C1 are computed and presented in the first five columns of Table 8.7. The last column of Table 8.7 displays the calculated local priority scores (v_{i1}) for alternatives under C1, solved using the eigenvalue method described in Step 2. Additionally, the CR is computed as 0.0427 (within acceptable threshold), obtained using the equation in Step 3.

TABLE 8.7

Pairwise Comparison Matrix for Alternatives Under C1.

	A1	A2	A3	A4	Local Priority Score (v_{i1})
A1	1	9.0000	3.5140	2.6762	0.5296
A2	0.1111	1	0.1542	0.1365	0.0388
A3	0.2846	6.4860	1	0.5441	0.1741
A4	0.3737	7.3238	1.8378	1	0.2575

For the remaining criteria (C2 and C3), similar calculations are performed to construct the pairwise comparison matrices for alternatives under each of them and compute the corresponding local priority scores (v_{i2} and v_{i3}) for alternatives, as presented in Tables 8.8 and 8.9, respectively; the CR values are found to be 0.0677 and 0.0389, respectively, both within the acceptable threshold.

TABLE 8.8

Pairwise Comparison Matrix for Alternatives Under C2.

	A1	A2	A3	A4	Local Priority Score (v_{i2})
A1	1	0.3121	3.6051	6.7958	0.2675
A2	3.2042	1	5.8093	9.0000	0.5831
A3	0.2774	0.1721	1	4.1907	0.1086
A4	0.1471	0.1111	0.2386	1	0.0409

TABLE 8.9
Pairwise Comparison Matrix for Alternatives Under C3.

	A1	A2	A3	A4	Local Priority Score (v_{i3})
A1	1	0.3245	0.1171	0.1111	0.0407
A2	3.0812	1	0.1548	0.1445	0.0825
A3	8.5392	6.4580	1	0.6846	0.3908
A4	9.0000	6.9188	1.4608	1	0.4859

Step 5. Aggregate the local priority scores using the criteria weights calculated in Step 2 to determine the global priority scores (P_i) for alternatives, as follows:

$$P_i = \sum_{j=1}^{n} w_j v_{ij}, \quad i \in \{1,2,\ldots,m\} \tag{8.16}$$

The alternatives are ranked based on the P_i values, where the alternative with the highest P_i value is top-ranked.

NUMERICAL CALCULATIONS

For instance, for $i = 1$ and using the criteria weights computed in Step 2:

$$P_1 = \sum_{j=1}^{3} w_j v_{1j} = 0.6267 \times 0.5296 + 0.2797 \times 0.2675 + 0.09 \times 0.0936 = 0.4105$$

Similarly, the performance scores for all alternatives are calculated as $P_1 = 0.4105$, $P_2 = 0.1951$, $P_3 = 0.1761$ and $P_4 = 0.2183$. Accordingly, the ranking is A1 ≻ A4 ≻ A2 ≻ A3, with A1 being the top-ranked alternative by AHP.

Alternatively, we can utilize SuperDecisions, a specialized software that has AHP and ANP implemented, to assist us in solving the foregoing numerical example. SuperDecisions (downloadable from https://superdecisions.com/) is a free educational software developed by the team of the creator of the methods, Professor Thomas Saaty. To begin, we construct the AHP model with the three hierarchical levels (namely, L1 – goal, L2 – criteria

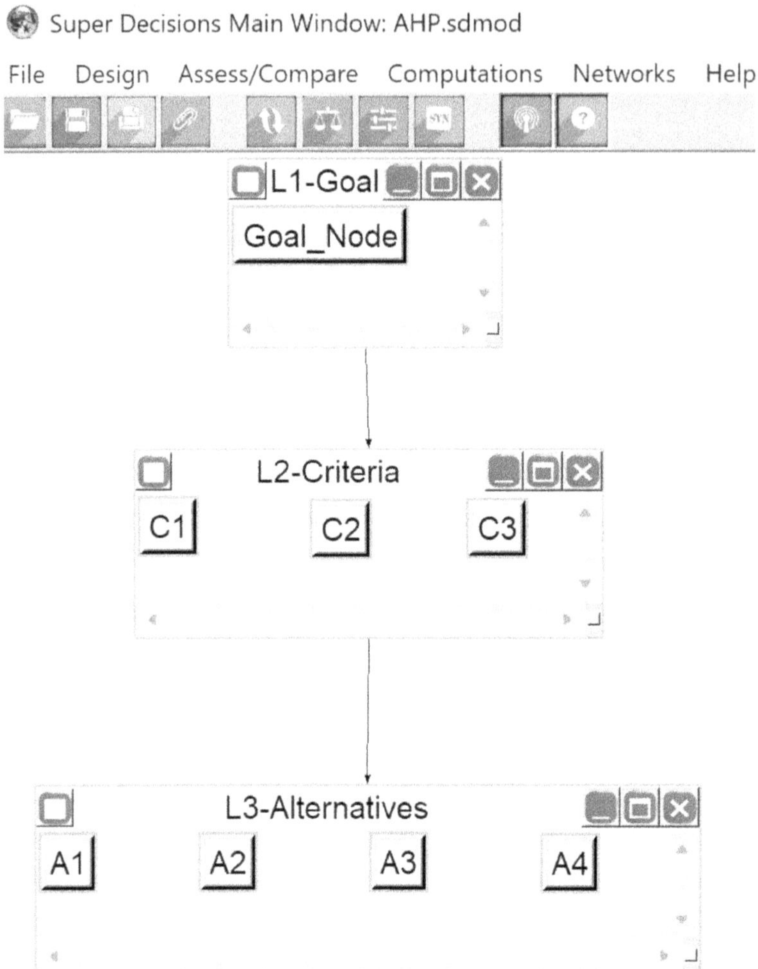

FIGURE 8.2(A)
AHP representation in SuperDecisions software.

and L3 – alternatives) within the interface of the software, as illustrated in Figure 8.2(a). Next, we input the pairwise comparison matrix for the three criteria (from the numerical calculations part of Step 1) and the pairwise comparison matrices for the four alternatives under each of the three criteria (from the first five columns of Tables 8.7, 8.8 and 8.9) into SuperDecisions.

The software then calculates the criteria weights and the local priorities of alternatives under each criterion. These results, highlighted in the red boxes of Figure 8.2(b), match the values we previously computed manually.

Super Decisions Main Window: AHP.sdmod: Unweighted Super Matrix

Cluster Node Labels		L1-Goal	L2-Criteria			L3-Alternatives			
		Goal_Node	C1	C2	C3	A1	A2	A3	A4
L1-Goal	Goal_Node	0.000000	0.000000	0.000000	0.000000	0.000000	0.000000	0.000000	0.000000
L2-Criteria	C1	0.626696	0.000000	0.000000	0.000000	0.000000	0.000000	0.000000	0.000000
	C2	0.279688	0.000000	0.000000	0.000000	0.000000	0.000000	0.000000	0.000000
	C3	0.093616	0.000000	0.000000	0.000000	0.000000	0.000000	0.000000	0.000000
L3-Alternatives	A1	0.000000	0.529640	0.267456	0.040746	0.000000	0.000000	0.000000	0.000000
	A2	0.000000	0.038790	0.583045	0.082529	0.000000	0.000000	0.000000	0.000000
	A3	0.000000	0.174103	0.108638	0.390851	0.000000	0.000000	0.000000	0.000000
	A4	0.000000	0.257467	0.040862	0.485875	0.000000	0.000000	0.000000	0.000000

Done

FIGURE 8.2 (B)
Criteria weights and local priorities of alternatives for each criterion inside red boxes; a screenshot from SuperDecisions for AHP.

Readers can ignore all the other values reported in Figure 8.2(b) for now; they will understand their meanings in the subsequent ANP section. Finally, SuperDecisions determines the global priority scores of the alternatives, presented in Figure 8.2(c), which also align with our previous calculations. The original SuperDecisions file for this AHP model can be downloaded from our shared folder in Google Drive:

https://drive.google.com/drive/folders/18ynIwnKaCRsptwBCrMIFVOlV9Mu_A7qM

Additional Comment: If an ACM based on objective measurement data is available (e.g., Table 8.1), apart from what Step 4 describes for mapping objective data into Saaty's 1–9 scale either through subjectively assessing the ACM or mathematical transformation, another straightforward approach is to directly utilize the ACM as is and adopt a hybrid approach combining AHP with other MCDM methods (e.g., SAW and MEW) for ranking alternatives. In this hybrid approach, AHP is used only to determine criteria weights subjectively, while the objective ACM provides the evaluation data for alternatives, which are then ranked using the MCDM method (e.g., SAW and MEW). This approach, also commonly adopted in the literature (Zaidan et al., 2015; Suartini et al., 2023), bypasses the need for mapping onto Saaty's 1–9 scale and leverages objective quantitative values directly.

The advantages of AHP are as follows: (1) It provides a systematic approach to decompose complex decision problems into hierarchical levels, beginning from the overall goal, moving down to criteria and finally to the alternatives.

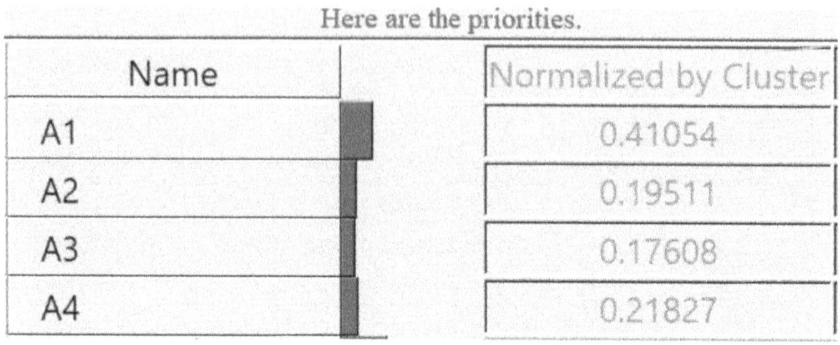
Here are the priorities.

Name	Normalized by Cluster
A1	0.41054
A2	0.19511
A3	0.17608
A4	0.21827

FIGURE 8.2(C)
Global priorities of alternatives; a screenshot from SuperDecisions for AHP.

(2) The use of pairwise comparisons allows subjective and qualitative judgments to be included explicitly in the decision-making process while preserving a measure of logical consistency through the CR. (3) By breaking down the comparisons at each hierarchical level, AHP helps the decision-maker focus on smaller, more manageable pairs, which can reduce cognitive overload compared to simultaneously considering all criteria and alternatives. (3) Due to its intuitive nature and well-established methodology, AHP is widely adopted and/or hybridized with other methods across a broad range of decision-making contexts, including engineering design, project selection, policymaking and resource allocation.

The limitations of AHP are as follows: (1) It heavily depends on the subjective judgments of decision-makers, which can introduce biases or inconsistencies, especially when dealing with large comparison matrices. (2) The use of Saaty's 1–9 scale, while straightforward, can reduce fidelity for nuanced differences; mapping a more granular measurement scale onto these discrete intervals may oversimplify actual distinctions among criteria or alternatives. (3) Rank reversal (a common limitation of many MCDM methods) can occur if new alternatives are added or existing ones are removed, since this can alter the relative priorities derived from the pairwise comparison matrices of alternatives. (4) As the number of criteria and alternatives grows, the required pairwise comparisons increase exponentially, resulting in greater computational effort and higher cognitive demand on the part of decision-makers.

8.5 Analytic Network Process (ANP)

The ANP method, developed by Saaty (1996, 1999), is a generalization and an advanced version of the AHP method described in Section 8.4. This more advanced method emphasizes that everything is interconnected to everything else, with a continuous flow of influence among all elements. ANP accommodates complex decision-making scenarios involving interdependencies and feedback among criteria and alternatives. Unlike AHP, which structures decision problems strictly into hierarchical layers with unidirectional influences, ANP uses networks, allowing mutual dependence among elements and bidirectional relationships. Therefore, ANP effectively addresses decision-making complexities, such as criteria influencing alternatives and alternatives simultaneously impacting the criteria themselves.

The fundamental difference between AHP and ANP is that the latter replaces the strict hierarchical structure with a more generalized network structure comprising clusters (e.g., criteria clusters and alternatives clusters) and nodes (e.g., individual criterion and alternative). These clusters and nodes can have inner dependencies (within the same cluster) and outer dependencies (across different clusters), creating loops and feedback. The steps for applying ANP are detailed as follows:

Step 1. Construct the ANP network structure. For clarity of illustration, we will use a simplified example to walk through the ANP process. Suppose a company needs to select one of the two alternative designs (A_1 and A_2) based on three criteria (C_1: cost; C_2: technical feasibility; and C_3: market risk). Decision-makers recognize some interdependence among these criteria and feedback from the alternatives to the criteria.

Thus, we have the following nodes:

- $\{G\}$ – single node for the goal in the goal cluster.
- $\{C_1, C_2, C_3\}$ – three criteria nodes in the criteria cluster.
- $\{A_1, A_2\}$ – two alternative nodes in the alternatives cluster.

Suppose the network of influences, as illustrated in Figure 8.3, is as follows:

- G influences all criteria (C_1, C_2, C_3); this does not differ from that of AHP.
- C_1 influences $\{C_2, C_3\}$ (e.g., a higher cost can correlate with higher technical feasibility and lower market risk).
  - C_2 influences $\{C_1, C_3\}$
  - C_3 influences $\{C_1, C_2\}$
- Each criterion (C_1, C_2, C_3) influences the alternatives (A_1, A_2); this does not differ from that of AHP.

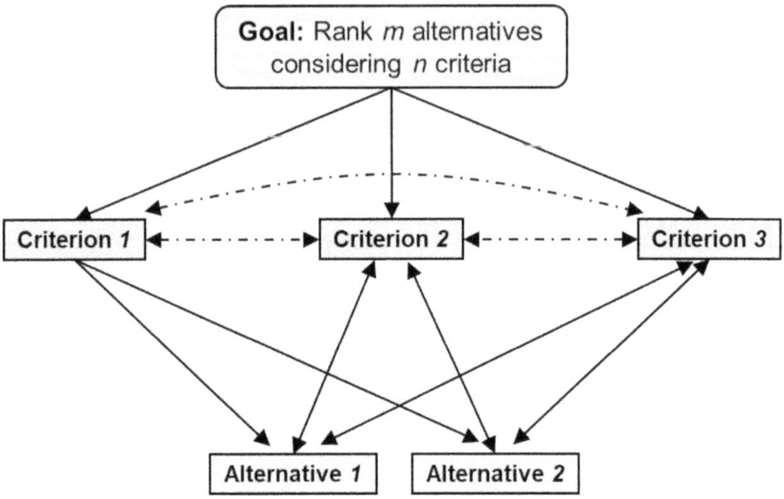

FIGURE 8.3
ANP representation for the simplified decision-making problem; dashed arrows indicate inner dependencies within the criteria cluster; and solid arrows represent outer dependencies across the goal, criteria and alternatives clusters. Note that some arrows are bidirectional, whereas the others (e.g., from criterion 1 to alternatives 1 and 2) are in one direction only.

- Each alternative (A_1, A_2) influences both C_2 and C_3 in a feedback loop. For example, each alternative, if implemented, might affect technical feasibility and market risk in the longer term. For instance, a more innovative design A_1 may push the company to adopt advanced technology solutions, raising the overall technical feasibility in the long run and simultaneously reduce future market risk by aligning the product with emerging trends.
- If a node influences both the criteria and alternatives clusters, we assume its impact on the alternatives cluster is always three times as important as its impact on the criteria cluster. In other words, the particular node devotes 75% of its influence to the alternatives and 25% to the criteria.

Step 2. Construct pairwise comparison matrices among the nodes and determine local priorities.

Using the simplified example, we now go node by node, identifying each parent node and performing the standard pairwise comparisons (as in AHP) among the child nodes under each parent to find how strongly each parent influences its child nodes.

8.5.1 Under the Goal Node (G)

Since G influences all criteria (C_1, C_2, C_3):

- Parent: G
- Children: $\{C_1, C_2, C_3\}$

Suppose the subjective assessment using Saaty's 1–9 scale is as follows:

- C_1 and C_2 are treated equally important (ratio 1:1)
- C_3 is twice as important as C_1 and also twice as important as C_2 (ratios 2:1)

Hence, the 3×3 pairwise comparison matrix of the three criteria under G is constructed as follows:

$$M_{G \to (C_1, C_2, C_3)} = \begin{pmatrix} 1 & 1 & 0.5 \\ 1 & 1 & 0.5 \\ 2 & 2 & 1.0 \end{pmatrix}$$

Same as the calculations in Step 2 of AHP, the eigenvalue method is employed to determine the principal eigenvalue ($\lambda_{\max}$) and the normalized principal eigenvector (i.e., criteria priorities or weights). Using WolframAlpha, $\lambda_{\max}$ is found to be 3.0. The corresponding criteria weights are as follows:

$$w_{G \to C_1} = 0.25,\ w_{G \to C_2} = 0.25,\ w_{G \to C_3} = 0.50.$$

For brevity, the CR check is omitted here. Its computation follows the exact same procedure as in Step 3 of AHP.

8.5.2 Under the Criterion 1 Node (C_1)

Since C_1 influences both $\{C_2, C_3\}$ and (A_1, A_2):

- Parent: C_1
- Children: $\{C_2, C_3, A_1, A_2\}$

Besides, as aforementioned, when a node influences both the criteria and alternatives clusters, it devotes 75% of its influence to the alternatives and 25% to the criteria.

Now, inside $\{C_2, C_3\}$ pairwise comparison, C_2 is one-half as important as C_3 (ratio 1:2). The comparison matrix is as follows:

$$M_{C_1 \to (C_2, C_3)} = \begin{pmatrix} 1 & 1/2 \\ 2 & 1 \end{pmatrix}$$

For this matrix, $\lambda_{\max} = 2$. The corresponding priorities are 0.3333 and 0.6667 for C_2 and C_3, respectively. Given that C_1 allocates 25% of its influence to the criteria cluster, the weights are:

$$w_{C_1 \to C_2} = 0.3333 \times 0.25 = 0.0833$$

$$w_{C_1 \to C_3} = 0.6667 \times 0.25 = 0.1667$$

Analogously, inside $\{A_1, A_2\}$ pairwise comparison, A_1 is three times as important as A_2 (ratio 3:1). The comparison matrix is as follows:

$$M_{C_1 \to (A_1, A_2)} = \begin{pmatrix} 1 & 3 \\ 1/3 & 1 \end{pmatrix}$$

For this matrix, $\lambda_{\max} = 2$. The corresponding priorities are 0.75 and 0.25 for A_1 and A_2, respectively. Given that C_1 allocates 75% of its influence to the alternatives cluster, the weights are:

$$w_{C_1 \to A_1} = 0.75 \times 0.75 = 0.5625$$

$$w_{C_1 \to A_2} = 0.25 \times 0.75 = 0.1875$$

8.5.3 Under the Criterion 2 Node (C_2)

The calculations under C_2 are similar to those under C_1; for brevity, only the key points and numbers are presented.

C_2 influences both $\{C_1, C_3\}$ and (A_1, A_2):

- Parent: C_2
- Children: $\{C_1, C_3, A_1, A_2\}$

Inside $\{C_1, C_3\}$, C_1 is treated equally as important as C_3 (ratio 1:1). The corresponding priorities are 0.25 each. Considering 25% of parents' influence on the criteria cluster, the weights are:

$$w_{C_2 \to C_1} = 0.25 \times 0.25 = 0.125$$

$$w_{C_2 \to C_3} = 0.25 \times 0.25 = 0.125$$

Inside $\{A_1, A_2\}$, A_1 is one-half as important as A_2 (ratio 1:2). The corresponding priorities are 0.3333 and 0.6667 for A_1 and A_2, respectively. Considering 75% of parents' influence on the alternatives cluster, the weights are:

$$w_{C_2 \to A_1} = 0.3333 \times 0.75 = 0.25$$

$$w_{C_2 \to A_2} = 0.6667 \times 0.75 = 0.50$$

8.5.4 Under the Criterion 3 Node (C_3)

Again, for conciseness, only the key points and numbers under C_3 are presented.

- Parent: C_3
- Children: $\{C_1, C_2, A_1, A_2\}$.

Inside $\{C_1, C_2\}$, C_1 is three times as important as C_3 (ratio 3:1). The corresponding priorities are 0.75 and 0.25 for C_1 and C_2. Considering 25% of parents' influence on the criteria cluster, the weights are:

$$w_{C_3 \to C_1} = 0.75 \times 0.25 = 0.1875$$

$$w_{C_3 \to C_2} = 0.25 \times 0.25 = 0.0625$$

Inside $\{A_1, A_2\}$, A_1 is twice as important as A_2 (ratio 2:1). The corresponding priorities are 0.6667 and 0.3333 for A_1 and A_2, respectively. Considering 75% of parents' influence on the alternatives cluster, the weights are:

$$w_{C_3 \to A_1} = 0.6667 \times 0.75 = 0.50$$

$$w_{C_3 \to A_2} = 0.3333 \times 0.75 = 0.25$$

8.5.5 Under the Alternative 1 Node (A_1)

Since A_1 influences both C_2 and C_3 in a feedback loop:

- Parent: A_1
- Children: $\{C_2, C_3\}$.

Inside $\{C_1, C_2\}$, suppose the ratio of $C_2 : C_3 = 1:2$, thus:

$$w_{A_1 \to C_2} = 0.3333$$

$$w_{A_1 \to C_3} = 0.6667$$

8.5.6 Under the Alternative 2 Node (A_2)

Since A_2 influences both C_2 and C_3 in a feedback loop:

- Parent: A_2
- Children: $\{C_2, C_3\}$.

Inside $\{C_1, C_2\}$, suppose the ratio of $C_2 : C_3 = 1:1$, thus:

$$w_{A_2 \to C_2} = 0.50$$

$$w_{A_2 \to C_3} = 0.50$$

Step 3. Construct the weighted supermatrix ($\mathbf{W}$).

The weighted supermatrix is formed by consolidating all the results from the previous step. It is termed "super" because it incorporates all the nodes and their influences in the ANP network. In this supermatrix, as shown in Table 8.10, each column header represents a parent node; the entries in each column indicate the influence of the parent node on its child nodes; for node pairs that were not compared in the previous step, simply assign a value of 0.

Step 4. Compute the limit supermatrix ($\mathbf{W}^\infty$) as follows:

$$\mathbf{W}^\infty = \lim_{k \to \infty} \mathbf{W}^k \tag{8.17}$$

As seen, the limit supermatrix is obtained by raising the weighted supermatrix to powers until convergence. This can be computed by repeatedly multiplying the weighted supermatrix by itself until it stabilizes. In essence, the limit supermatrix represents the long-run steady-state priorities in a network of feedback influences. Each column of the supermatrix tells us how

TABLE 8.10

ANP Weighted Supermatrix ($\mathbf{W}$).

	G	C_1	C_2	C_3	A_1	A_2
G	0	0	0	0	0	0
C_1	0.25	0	0.125	0.1875	0	0
C_2	0.25	0.0833	0	0.0625	0.3333	0.50
C_3	0.50	0.1667	0.125	0	0.6667	0.50
A_1	0	0.5625	0.25	0.50	0	0
A_2	0	0.1875	0.50	0.25	0	0

TABLE 8.11

ANP Limit Supermatrix ($\mathbf{W}^{\infty}$).

	G	C_1	C_2	C_3	A_1
G	0	0	0	0	0
C_1	0.0797	0.0797	0.0797	0.0797	0.0797
C_2	0.1991	0.1991	0.1991	0.1991	0.1991
C_3	0.2926	0.2926	0.2926	0.2926	0.2926
A_1	0.2409	0.2409	0.2409	0.2409	0.2409
A_2	0.1876	0.1876	0.1876	0.1876	0.1876

a particular parent node distributes its influence among its child nodes. By raising the supermatrix to higher and higher powers, it is essentially simulating infinitely many rounds of mutual influence among the nodes. Once the matrix powers stabilize (converge), it indicates a steady-state distribution of influence. The entries in the columns of the resulting $\mathbf{W}^{\infty}$, as shown in Table 8.11, can be interpreted as the overall priorities of each node under its parent node. In this example, all columns eventually converge to the same values; however, this may not be the case in all applications.

Next, from $\mathbf{W}^{\infty}$, we extract the global priority scores for alternatives, $A_1 : 0.2409$ and $A_2 : 0.1876$, under the goal node G. After applying sum normalization, the normalized global priority score for A_1 is $0.5622(= 0.2409/(0.2409 + 0.1876)$ and for A_2 is $0.4378(= 0.1876/(0.2409 + 0.1876)$. Therefore, the alternative A_1 is ranked higher and recommended for implementation.

Alternatively, we can also use SuperDecisions to solve this problem. Figure 8.4(a) presents the weighted supermatrix $(\mathbf{W})$ calculated by SuperDecisions. Figure 8.4(b) displays the limit supermatrix $(\mathbf{W}^{\infty})$. Figure 8.4(c) shows the normalized global priority scores. All these results tally with our previous calculations. The original SuperDecisions file for this ANP model can be downloaded from our shared folder in Google Drive:

https://drive.google.com/drive/folders/18ynIwnKaCRsptwBCrMIFVOlV9Mu_A7qM

Super Decisions Main Window: Chapter_8_ANP.sdmod...

Cluster Node Labels		L1-GOAL	L2-Criteria			L3-Alternatives	
		Goal_Node	C1	C2	C3	A1	A2
L1-GOAL	Goal_Node	0.000000	0.000000	0.000000	0.000000	0.000000	0.000000
L2-Criteria	C1	0.250000	0.000000	0.125000	0.187500	0.000000	0.000000
	C2	0.250000	0.083333	0.000000	0.062500	0.333333	0.500000
	C3	0.500000	0.166667	0.125000	0.000000	0.666667	0.500000
L3-Alternatives	A1	0.000000	0.562500	0.250000	0.500000	0.000000	0.000000
	A2	0.000000	0.187500	0.500000	0.250000	0.000000	0.000000

FIGURE 8.4(A)
Screenshot of the weighted supermatrix ($\mathbf{W}$) from SuperDecisions for ANP.

Super Decisions Main Window: Chapter_8_ANP.sdmod...

Cluster Node Labels		L1-GOAL	L2-Criteria			L3-Alternatives	
		Goal_Node	C1	C2	C3	A1	A2
L1-GOAL	Goal_Node	0.000000	0.000000	0.000000	0.000000	0.000000	0.000000
L2-Criteria	C1	0.079748	0.079748	0.079748	0.079748	0.079748	0.079748
	C2	0.199064	0.199064	0.199064	0.199064	0.199064	0.199064
	C3	0.292616	0.292616	0.292616	0.292616	0.292616	0.292616
L3-Alternatives	A1	0.240932	0.240932	0.240932	0.240932	0.240932	0.240932
	A2	0.187639	0.187639	0.187639	0.187639	0.187639	0.187639

Done

FIGURE 8.4(B)
Screenshot of the limit supermatrix ($\mathbf{W}^{\infty}$) from SuperDecisions for ANP.

The advantages of ANP are as follows: (1) ANP is designed to handle interdependent criteria and alternatives by allowing feedback loops and reciprocal influences, thereby providing a more realistic representation of complex decision problems that are not strictly hierarchical. (2) The network-based model facilitates a deeper understanding of how changes to one criterion can impact others, leading to comprehensive insights that might not emerge from linear approaches. (3) By breaking down interrelationships into pairwise comparisons,

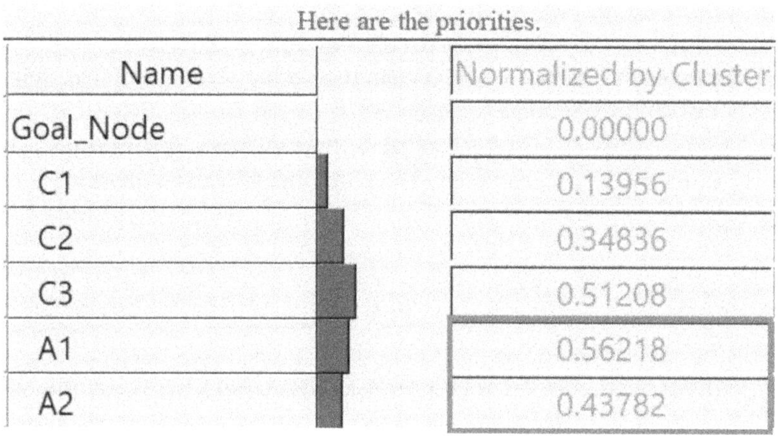

Here are the priorities.

Name	Normalized by Cluster
Goal_Node	0.00000
C1	0.13956
C2	0.34836
C3	0.51208
A1	0.56218
A2	0.43782

FIGURE 8.4(C)
Screenshot of the normalized global priority scores from SuperDecisions for ANP.

ANP uses subjective judgments while preserving logical consistency to a reasonable extent. (4) Because it captures the nuanced interplay among factors, ANP can be particularly beneficial in situations where multiple variables are intertwined and traditional methods struggle to capture those interdependencies.

The limitations of ANP are as follows: (1) ANP's feedback-driven structure can be challenging to explain to stakeholders who are unfamiliar with the method, potentially hindering consensus or agreement. (2) The method generally requires specialized software (e.g., SuperDecisions and the AhpAnpLib Python library) and a solid grasp of the underlying theory, which can be a barrier in terms of both cost and the learning curve for practitioners. (3) Verification and interpretation of results can be complex, given the myriad loops and interconnections; isolating the impact of individual criteria becomes less straightforward than in hierarchical models. (4) For simpler decision problems where criteria and alternatives are relatively independent, ANP may be excessive, since more familiar and less resource-intensive methods (e.g., AHP) can often suffice while providing similar outcomes.

8.6 Complex Proportional Assessment (COPRAS)

The COPRAS method, proposed by Zavadskas et al. (1994), is another aggregation-type MCDM method. It employs a systematic approach to rank alternatives based on their relative importance. After constructing the weighted normalized ACM, COPRAS calculates the sum of weighted normalized values for all maximization criteria and the sum of weighted normalized values for all minimization criteria for each alternative. These two sums are then aggregated using the method's specific equations (Zavadskas et al., 1994) to determine each alternative's performance score. The alternative with the highest performance score is top-ranked.

Step 1. Normalize the original ACM with m rows (i.e., alternatives) and n columns (i.e., criteria) using the sum normalization method.

$$F_{ij} = \frac{f_{ij}}{\sum_{k=1}^{m} f_{kj}} \tag{8.18}$$

NUMERICAL CALCULATIONS (USING ACM FROM TABLE 8.1)

For instance, for $i = 1$ and $j = 2$:

$$F_{12} = \frac{f_{12}}{\sum_{k=1}^{m} f_{k2}} = \frac{600}{600 + 700 + 500 + 400} = 0.2727$$

Similar calculations are performed for all other values in the ACM across the four alternatives and three criteria. The complete results, forming the normalized ACM, are shown in Table 8.12.

TABLE 8.12

Normalized ACM for COPRAS Walk-Through.

Alternatives	C1	C2	C3
A1	0.3069	0.2727	0.3982
A2	0.1683	0.3182	0.3055
A3	0.2541	0.2273	0.1525
A4	0.2706	0.1818	0.1438

Step 2. Construct the weighted normalized ACM by multiplying the values of each criterion by its assigned weight, w_j.

$$v_{ij} = F_{ij} \times w_j \tag{8.19}$$

NUMERICAL CALCULATIONS

For instance, for $i = 1$ and $j = 2$:

$$v_{12} = F_{12} \times w_2 = 0.2727 \times 0.33 = 0.0900$$

Likewise, calculations are carried out for all other values in the normalized ACM, using their respective assigned weights. The results, which constitute the weighted normalized ACM, are displayed in Table 8.13.

TABLE 8.13

Weighted Normalized ACM for COPRAS Walk-Through.

Alternatives	C1	C2	C3
A1	0.0767	0.0900	0.1672
A2	0.0421	0.1050	0.1283
A3	0.0635	0.0750	0.0641
A4	0.0677	0.0600	0.0604

Step 3. For each alternative, calculate the sums of weighted normalized values for both maximization and minimization criteria, S_{i+} and S_{i-}, separately.

$$S_{i+} = \sum_{j=1}^{g} v_{ij} \tag{8.20}$$

$$S_{i-} = \sum_{j=g+1}^{n} v_{ij} \tag{8.21}$$

Here, g is the number of maximization criteria, which can be arranged in the first g columns of the weighted normalized ACM. The minimization criteria are in columns $g+1$ to n. If all criteria are for maximization, then $g=n$; whereas $g=0$ when all criteria are for minimization.

NUMERICAL CALCULATIONS

There are two maximization criteria, hence $g=2$.

For instance, for $i=1$:

$$S_{1+} = \sum_{j=1}^{2} v_{1j} = v_{11} + v_{12} = 0.0767 + 0.0900 = 0.1667$$

$$S_{1-} = \sum_{j=2+1}^{3} v_{1j} = v_{13} = 0.1672$$

After calculating the S_{i+} and S_{i-} of each alternative, the results are tabulated in the first three columns of Table 8.14.

TABLE 8.14

Weighted Normalized ACM for COPRAS Walk-Through.

Alternatives	S_{i+}	S_{i-}	P_i
A1	0.1667	0.1672	0.2214
A2	0.1471	0.1283	0.2183
A3	0.1385	0.0641	0.2813
A4	0.1277	0.0604	0.2790

Step 4. Compute the performance score (P_i) of each alternative as follows:

$$P_i = \begin{cases} S_{i+} + \dfrac{\sum_{k=1}^{m} S_{k-}}{S_{i-}\sum_{k=1}^{m}\dfrac{1}{S_{k-}}}, & \text{if both maximization and minimization criteria exist} \\ S_{i+}, & \text{if only maximization criteria exist} \\ \dfrac{\sum_{k=1}^{m} S_{k-}}{S_{i-}\sum_{k=1}^{m}\dfrac{1}{S_{k-}}}, & \text{if only minimization criteria exist} \end{cases} \quad (8.22)$$

In the above equation, both $\sum_{k=1}^{m} S_{k-}$ and $\sum_{k=1}^{m} \frac{1}{S_{k-}}$ are invariant across the performance score (P_i) calculations for each alternative i and so the term $\frac{\sum_{k=1}^{m} S_{k-}}{S_{i-}\sum_{k=1}^{m}\frac{1}{S_{k-}}}$ can be considered as a scaled S_{i-} value. Further, a smaller S_{i-} leads to a higher P_i value. The alternative with the largest P_i is top-ranked and recommended to the decision-maker.

NUMERICAL CALCULATIONS

For instance, for $i = 1$, since both maximization and minimization criteria exist:

$$P_1 = S_{1+} + \frac{\sum_{k=1}^{m} S_{k-}}{S_{1-}\sum_{k=1}^{m}\frac{1}{S_{k-}}} = 0.1667$$

$$+ \frac{0.1672 + 0.1283 + 0.0641 + 0.0604}{0.1672\left(\frac{1}{0.1672} + \frac{1}{0.1283} + \frac{1}{0.0641} + \frac{1}{0.0604}\right)} = 0.2214$$

Similarly, performance scores for all alternatives are calculated and given in the last column of Table 8.14. Accordingly, the ranking is A3 > A4 > A1 > A2, with A3 being the top-ranked alternative by COPRAS.

The advantages of COPRAS are as follows: (1) COPRAS explicitly separates benefit (to be maximized) and cost (to be minimized) criteria, making it intuitive for decision-makers to see how each criterion contributes to or detracts from the final performance score of an alternative. (2) COPRAS has been successfully adopted in many applications (Stefano et al., 2015).

The limitations of COPRAS are as follows: (1) Because the final performance score is largely driven by the sum of weighted normalized values, exceptionally high (or low) performance on one criterion may greatly overshadow moderate performance on other criteria, which may not be desirable in some applications. (2) As with other MCDM methods, the addition or removal of alternatives may cause rank reversal.

8.7 Multi-Objective Optimization on the Basis of Ratio Analysis (MOORA)

The MOORA method, introduced by Brauers and Zavadskas (2006), is extensively applied in various fields for MCDM. After constructing the weighted normalized ACM, the performance score for each alternative is calculated by subtracting the aggregate of minimization criteria values from the aggregate of maximization criteria values. The alternative that has a larger performance score is ranked higher. The MOORA method follows a structured sequence of the following steps.

Step 1. Normalize the original ACM with m rows (i.e., alternatives) and n columns (i.e., criteria) by the vector normalization method.

$$F_{ij} = \frac{f_{ij}}{\sqrt{\sum_{k=1}^{m} f_{kj}^{2}}} \tag{8.23}$$

NUMERICAL CALCULATIONS (USING ACM FROM TABLE 8.1)

For instance, for $i = 1$ and $j = 2$:

$$F_{12} = \frac{f_{12}}{\sqrt{\sum_{k=1}^{m} f_{k2}^{2}}} = \frac{600}{\sqrt{600^2 + 700^2 + 500^2 + 400^2}} = 0.5345$$

Similar calculations are performed for all other values in the ACM across the four alternatives and three criteria. The complete results, forming the normalized ACM, are shown in Table 8.15.

TABLE 8.15

Normalized ACM for MOORA Walk-Through.

Alternatives	C1	C2	C3
A1	0.6015	0.5345	0.7321
A2	0.3299	0.6236	0.5617
A3	0.4980	0.4454	0.2804
A4	0.5304	0.3563	0.2644

Step 2. Construct the weighted normalized ACM by multiplying the values of each criterion by its assigned weight, w_j.

$$v_{ij} = F_{ij} \times w_j \tag{8.24}$$

NUMERICAL CALCULATIONS

For instance, for $i = 1$ and $j = 2$:

$$v_{12} = F_{12} \times w_2 = 0.5345 \times 0.33 = 0.1764$$

Likewise, calculations are carried out for all other values in the normalized ACM, using their respective assigned weights. The results, which constitute the weighted normalized ACM, are displayed in Table 8.16.

TABLE 8.16

Weighted Normalized ACM for MOORA Walk-Through.

Alternatives	C1	C2	C3
A1	0.1504	0.1764	0.3075
A2	0.0825	0.2058	0.2359
A3	0.1245	0.1470	0.1178
A4	0.1326	0.1176	0.1111

Step 3. Compute the performance score (P_i) of each alternative as follows:

$$P_i = \sum_{j=1}^{g} v_{ij} - \sum_{j=g+1}^{n} v_{ij} \tag{8.25}$$

Here, the significance of g is the same as that in the COPRAS method, meaning that g is the number of maximization criteria, which can be arranged in the first g columns of the weighted normalized ACM. The minimization criteria are in columns $g+1$ to n. The alternative with the largest P_i is top-ranked and recommended to the decision-maker. The performance score in Eq. 8.25 is somewhat simpler than that in the COPRAS method (Eq. 8.22).

NUMERICAL CALCULATIONS

There are two maximization criteria, hence $g = 2$. For instance, for $i = 1$:

$$P_1 = \sum_{j=1}^{2} v_{1j} - \sum_{j=2+1}^{3} v_{1j} = 0.1504 + 0.1764 - 0.3075 = 0.0193$$

Similarly, the performance scores for all alternatives are calculated as $P_1 = 0.0193$, $P_2 = 0.0523$, $P_3 = 0.1537$ and $P_4 = 0.1391$. Accordingly, the ranking is A3 ≻ A4 ≻ A2 ≻ A1, with A3 being the top-ranked alternative by MOORA. Note that some or all P_i can be negative (e.g., if all/majority of criteria are minimization type).

The advantages of MOORA are as follows: (1) It employs a straightforward mathematical formulation where the performance of each alternative is determined by summing weighted benefit criteria and then subtracting weighted cost criteria, making it easy to interpret. (2) The steps are relatively simple, which allows for efficient implementation across a wide variety of decision-making contexts.

The limitations of MOORA are as follows: (1) Similar to COPRAS, a single high or low value in one criterion can disproportionately influence the final performance score, which may be undesirable for decisions requiring a balanced performance across all criteria. (2) As with other MCDM methods, the addition or removal of alternatives may cause rank reversal.

8.8 Faire Un Choix Adéquat (FUCA)

FUCA is the French acronym for "Faire Un Choix Adéquat", which translates to "Make an Adequate Choice". The FUCA method, proposed by Fernando et al. (2011), determines the best alternative based on rank aggregation. It works by assigning ranks to alternatives with respect to each criterion and then computing a WS of these ranks. The alternative with the smallest aggregated rank is top-ranked and recommended to the decision-maker (Wang and Rangaiah, 2017). Thus, no normalization of criteria values is required.

Step 1. Rank the alternatives for each criterion. Each alternative is assigned a rank with respect to every criterion, denoted as r_{ij}, based on its performance under that criterion, as follows:

- Rank 1 is assigned to the best alternative.
- Rank 2 is assigned to the next best alternative.
- . . .
- Rank m (where m is the number of alternatives) is assigned to the worst alternative.

The ranking assignment depends on whether the criterion is of a benefit or cost type. For benefit criteria (maximization), the best alternative is the one with the highest value, receiving rank 1, while the worst receives rank m. Conversely, for cost criteria (minimization), the best alternative is the one with the lowest value, receiving rank 1, while the worst receives rank m.

NUMERICAL CALCULATIONS (USING ACM FROM TABLE 8.1)

For instance, for $i=2$ and $j=1$, by inspecting Table 8.1, we can easily find that the rank of A2 under C1 is $r_{21}=4$. Similar rankings are performed for all other values in the ACM across the four alternatives and three criteria. The complete results are shown in Table 8.17.

TABLE 8.17

Rank of Alternatives Under Each Criterion for FUCA Walk-Through.

Alternatives	C1	C2	C3
A1	1	2	4
A2	4	1	3
A3	3	3	2
A4	2	4	1

Step 2. Compute the aggregated ranks. Once the rankings under each criterion are assigned, a weighted rank summation of all criteria (R_i) is performed for each alternative.

$$R_i = \sum_{j=1}^{n} \left(r_{ij} \times w_j \right) \tag{8.26}$$

The alternative with the smallest R_i is top-ranked and recommended to the decision-maker.

NUMERICAL CALCULATIONS

For instance, for $i=4$:

$$R_4 = \sum_{j=1}^{3} \left(r_{4j} \times w_j \right) = 2\times 0.25 + 4\times 0.33 + 1\times 0.42 = 2.24$$

Similarly, the aggregated ranks for all alternatives are calculated as $R_1 = 2.59,$ $R_2 = 2.59,$ $R_3 = 2.58$ and $P_4 = 2.24$. Accordingly, the ranking is A4 ≻ A3 ≻ A1 = A2, with A4 being the top-ranked alternative by FUCA. Note that R_i can have the same value for two or more alternatives.

The advantages of FUCA are as follows: (1) It is easy to understand and implement, requiring only ranking assignments and simple weighted summation calculations. (2) Since FUCA ranks alternatives directly instead of normalizing numerical values for each criterion, it eliminates the dependence of ranking on the normalization method. This helps avoid issues such as possibly drastic changes in normalized values when ACM is updated. (3) FUCA is less sensitive to extreme values, as it considers only rankings and not absolute magnitudes.

The limitations of FUCA are as follows: (1) Since FUCA relies solely on ranks, it does not account for the actual differences between alternative performances. A small difference in criteria values may be treated the same as a large difference (i.e., in terms of ranking difference). (2) In cases where multiple alternatives have the same criterion value, additional tie-breaking rules must be defined, which can introduce subjectivity. (3) FUCA is still susceptible to rank reversal when new alternatives are introduced or existing ones are removed, as the reassignment of ranks for each criterion may alter the relative positioning of alternatives.

8.9 Weighted Aggregated Sum Product Assessment (WASPAS)

The WASPAS method, introduced in Zavadskas et al. (2012), is a hybrid MCDM approach that combines both the SAW and MEW methods in its algorithm. It calculates a composite performance score for each alternative by blending the additive results from SAW with the multiplicative results from MEW. By integrating these two different perspectives, WASPAS aims to leverage the strengths of each method while mitigating their individual shortcomings. The procedure of the WASPAS method consists of the following steps.

Step 1. Normalize the original ACM with m rows (i.e., alternatives) and n columns (i.e., criteria) using the max normalization. This step is the same as Step 1 of the SAW and MEW methods (Sections 8.2 and 8.3). For the sake of brevity, the normalization equations and numerical calculations (same as Table 8.2) are not repeated here.

Step 2. Calculate the performance score (P_i) for each alternative by integrating additive (as in the SAW method) and multiplicative (as in the MEW method) components, as follows: The alternative with the largest P_i is top-ranked and recommended to the decision-maker.

$$P_i = \lambda \sum_{j=1}^{n} F_{ij} w_j + (1-\lambda) \prod_{j=1}^{n} F_{ij}^{w_j} \tag{8.27}$$

Here, $\lambda \in [0,1]$ is a parameter that balances between the additive and multiplicative components. When $\lambda = 1$, WASPAS reduces to a purely additive approach (i.e., SAW). When $\lambda = 0$, it simplifies to a fully multiplicative approach (i.e., MEW). Intermediate λ values allow a flexible combination of both approaches. Commonly, $\lambda = 0.5$ is chosen to give equal priority to both SAW and MEW components.

NUMERICAL CALCULATIONS

For instance, for $i = 1$ and $\lambda = 0.5$, using the normalized criteria values in Table 8.2:

$$\begin{aligned} P_1 &= 0.5 \sum_{j=1}^{3} F_{1j} w_j + 0.5 \prod_{j=1}^{3} F_{1j}^{w_j} \\ &= 0.5(1 \times 0.25 + 0.8571 \times 0.33 + 0.3612 \times 0.42) \\ &+ 0.5\left(1^{0.25} \times 0.8571^{0.33} \times 0.3612^{0.42}\right) \\ &= 0.6521 \end{aligned}$$

Similarly, the performance scores for all alternatives are calculated as $P_1 = 0.6521$, $P_2 = 0.6460$, $P_3 = 0.8358$ and $P_4 = 0.8173$. Accordingly, the ranking is A3 > A4 > A1 > A2, with A3 being the top-ranked alternative by WASPAS.

The advantages of WASPAS are as follows: (1) It involves straightforward calculations, making it easy to implement and computationally efficient. (2) By integrating SAW and MEW, WASPAS takes advantage of each method's strengths: The SAW part helps maintain clarity and proportional scaling, while the MEW part ensures no single criterion is completely overshadowed by extremely high values in others.

The limitations of WASPAS are as follows: (1) The final ranking can be affected by the choice of λ value, which itself becomes an additional decision-making challenge. (2) If a criterion value is zero or extremely small for an alternative, the multiplicative component can excessively penalize it. (3) Similar to other MCDM methods, the addition or removal of alternatives can affect rankings, potentially leading to rank reversals.

8.10 Summary

Main points in this chapter on aggregation-type MCDM methods are as follows:

1. This chapter describes and illustrates eight aggregation-type MCDM methods, each employing distinct principles for aggregating criteria values to determine an overall performance score for alternatives. Listed chronologically, these methods are SAW, MEW, AHP, ANP, COPRAS, MOORA, FUCA and WASPAS.
2. Aggregation-type MCDM methods evaluate alternatives by combining criteria values and weights using additive, multiplicative, or hybrid aggregation techniques. These techniques transform and combine criteria values into a single overall performance score, allowing direct ranking of alternatives.
3. Each method in this chapter is systematically presented with a clear explanation of its theoretical foundation, step-by-step algorithm and numerical calculations.
4. The chapter discusses the advantages and limitations associated with each aggregation-type MCDM method, aiding decision-makers in identifying appropriate methods aligned with specific requirements and decision contexts.
5. The hierarchical and network-based MCDM methods (i.e., AHP and ANP) offer structured approaches for incorporating subjective expert judgments through pairwise comparisons, distinguishing them from other aggregation-type methods.
6. Rank reversal, where adding or removing alternatives might alter the relative rankings of alternatives, remains a notable limitation across MCDM methods including aggregation-type methods.
7. A thorough understanding of the principles and relative merits of these aggregation-type MCDM methods enables decision-makers to select an appropriate method or combination of methods tailored to the specific characteristics and requirements of a decision-making scenario.

8. Finally, for methods that use the common ACM (Table 8.1) to walk through their steps, namely, SAW, MEW, COPRAS, MOORA, FUCA and WASPAS, it is observed that, except for FUCA, all methods identify A3 as the top-ranked alternative. In contrast, FUCA ranks A4 as the top alternative. The ranking order of the remaining alternatives varies across methods, as detailed in their respective sections. These ranking results cannot be generalized and it is better to try several MCDM methods for any application.

References

Baydaş, M., Kavacık, M., & Wang, Z. (2024). Interpreting the determinants of sensitivity in MCDM methods with a new perspective: An application on e-scooter selection with the PROBID method. *Spectrum of Engineering and Management Sciences*, *2*(1), 17–35.

Brauers, W. K., & Zavadskas, E. K. (2006). The MOORA method and its application to privatization in a transition economy. *Control and Cybernetics*, *35*(2), 445–469.

Fernando, M. M. L., Escobedo, J. L. P., Azzaro-Pantel, C., Pibouleau, L., Domenech, S., & Aguilar-Lasserre, A. (2011, April). Selecting the best portfolio alternative from a hybrid multiobjective GA-MCDM approach for new product development in the pharmaceutical industry. In *2011 IEEE Symposium on Computational Intelligence in Multicriteria Decision-Making (MDCM)* (pp. 159–166). IEEE.

Karim, R., & Karmaker, C. L. (2016). Machine selection by AHP and TOPSIS methods. *American Journal of Industrial Engineering*, *4*(1), 7–13.

MacCrimmon, K. R. (1968). Decisionmaking among multiple-attribute alternatives: A survey and consolidated approach. *RAND Corporation Santa Monica*. https://www.rand.org/content/dam/rand/pubs/research_memoranda/2009/RM4823.pdf (accessed in March 2025).

Miller, D. W., & Starr, M. K. (1969). *Executive decision and operations research* (2nd ed.). Prentice Hall.

Nabavi, S. R., Jafari, M. J., & Wang, Z. (2023). Deep learning aided multi-objective optimization and multi-criteria decision making in thermal cracking process for olefines production. *Journal of the Taiwan Institute of Chemical Engineers*, *152*, 105179.

Nabavi, S. R., Wang, Z., & Rodríguez, M. L. (2024). Multi-objective optimization and multi-criteria decision-making approach to design a multi-tubular packed-bed membrane reactor in oxidative dehydrogenation of ethane. *Energy & Fuels*, *39*(1), 491–503.

Podvezko, V. (2009). Application of AHP technique. *Journal of Business Economics and Management*, (2), 181–189.

Saaty, T. L. (1977). A scaling method for priorities in hierarchical structures. *Journal of Mathematical Psychology*, *15*(3), 234–281.

Saaty, T. L. (1990). How to make a decision: The analytic hierarchy process. *European Journal of Operational Research*, *48*(1), 9–26.

Saaty, T. L. (1996). *Decisions with the Analytic Network Process (ANP)*. University of Pittsburgh (USA), ISAHP, 96.
Saaty, T. L. (1999). Fundamentals of the analytic network process. In *Proceedings of the 5th International Symposium on the Analytic Hierarchy Process* (Vol. 12, No. 14), Kobe, Japan.
Si, J., Marjanovic-Halburd, L., Nasiri, F., & Bell, S. (2016). Assessment of building-integrated green technologies: A review and case study on applications of Multi-Criteria Decision Making (MCDM) method. *Sustainable Cities and Society, 27*, 106–115.
Stefano, N. M., Casarotto Filho, N., Vergara, L. G. L., & da Rocha, R. U. G. (2015). COPRAS (Complex Proportional Assessment): State of the art research and its applications. *IEEE Latin America Transactions, 13*(12), 3899–3906.
Suartini, N. K. Y., Divayana, D. G. H., & Dewi, L. J. E. (2023). Comparison analysis of AHP-SAW, AHP-WP, AHP-TOPSIS methods in private tutor selection. *International Journal of Modern Education and Computer Science, 13*(1), 28.
Teknomo, K. (2006). *Analytic Hierarchy Process (AHP) tutorial*. http://people.revoledu.com/kardi/tutorial/AHP (accessed in March 2025).
Wang, Z., Li, J., Rangaiah, G. P., & Wu, Z. (2022). Machine learning aided multi-objective optimization and multi-criteria decision making: Framework and two applications in chemical engineering. *Computers & Chemical Engineering, 165*, 107945.
Wang, Z., Nabavi, S. R., & Rangaiah, G. P. (2024). Multi-criteria decision making in chemical and process engineering: Methods, progress and potential. *Processes, 12*(11), 2532.
Wang, Z., Parhi, S. S., Rangaiah, G. P., & Jana, A. K. (2020). Analysis of weighting and selection methods for Pareto-optimal solutions of multiobjective optimization in chemical engineering applications. *Industrial & Engineering Chemistry Research, 59*(33), 14850–14867.
Wang, Z., & Rangaiah, G. P. (2017). Application and analysis of methods for selecting an optimal solution from the Pareto-optimal front obtained by multiobjective optimization. *Industrial & Engineering Chemistry Research, 56*(2), 560–574.
Wang, Z., Tan, W. G. Y., Rangaiah, G. P., & Wu, Z. (2023). Machine learning aided model predictive control with multi-objective optimization and multi-criteria decision making. *Computers & Chemical Engineering, 179*, 108414.
Zaidan, A. A., Zaidan, B. B., Hussain, M., Haiqi, A., Kiah, M. M., & Abdulnabi, M. (2015). Multi-criteria analysis for OS-EMR software selection problem: A comparative study. *Decision Support Systems, 78*, 15–27.
Zavadskas, E. K., Kaklauskas, A., & Šarka, V. (1994). The new method of multicriteria complex proportional assessment of projects. *Technological and Economic Development of Economy, 1*, 131–139.
Zavadskas, E. K., Turskis, Z., Antucheviciene, J., & Zakarevicius, A. (2012). Optimization of weighted aggregated sum product assessment. *Elektronika ir elektrotechnika, 122*(6), 3–6.

Exercises

E8.1 You are given an ACM consisting of three alternatives (A1, A2 and A3) and four criteria (C1, C2, C3 and C4), with the following values:

Alternatives	C1	C2	C3	C4
A1	0.78	117	6.2	35
A2	0.65	150	5.8	40
A3	0.83	110	4.9	30

Criteria C1, C2 and C3 are benefit criteria to be maximized, whereas C4 is a cost criterion to be minimized. The weights for the criteria are $w_1 = 0.30$, $w_2 = 0.25$, $w_3 = 0.20$ and $w_4 = 0.25$. Apply all (or some of) the aggregation-type MCDM methods covered in this chapter to rank the alternatives. Identify the top-ranked alternative by each method and compare them.

E8.2 You are given an ACM consisting of four alternatives (A1, A2, A3 and A4) and five criteria (C1, C2, C3, C4 and C5), with the following values:

Alternatives	C1	C2	C3	C4	C5
A1	95	0.245	13.7	72.1	0.089
A2	88	0.312	11.9	69.5	0.075
A3	92	0.298	12.5	74.8	0.103
A4	85	0.331	14.1	70.3	0.091

Criteria C1, C2 and C4 are benefit criteria to be maximized, whereas C3 and C5 are cost criteria to be minimized. The weights for the criteria are $w_1 = 0.20$, $w_2 = 0.20$, $w_3 = 0.25$, $w_4 = 0.15$ and $w_5 = 0.20$. Apply all (or some of) the aggregation-type MCDM methods covered in this chapter to rank the alternatives. Identify the top-ranked alternative by each method and compare them.

E8.3 You are given an ACM consisting of six alternatives (A1, A2, A3, A4, A5 and A6) and three criteria (C1, C2 and C3), with the following values:

Alternatives	C1	C2	C3
A1	5400	8.1	0.120
A2	4750	7.5	0.135
A3	6100	9.0	0.110
A4	5900	8.3	0.095
A5	5150	8.8	0.145
A6	5650	7.9	0.105

Criteria C1 and C2 are benefit criteria to be maximized, whereas C3 is a cost criterion to be minimized. The weights for the criteria are $w_1 = 0.40$, $w_2 = 0.35$ and $w_3 = 0.25$. Apply all (or some of) the aggregation-type MCDM methods covered in this chapter to rank the alternatives. Identify the top-ranked alternative by each method and compare them.

E8.4 Describe the similarities and differences between the SAW and MEW methods.

E8.5 Describe the similarities and differences between the AHP and ANP methods.

E8.6 Describe the similarities and differences among the COPRAS, MOORA and FUCA methods.

E8.7 Explain the role of the λ parameter in WASPAS. Conduct sensitivity analysis by varying the value of λ from 0 to 1 in increments of 0.1 for the ACM provided in exercise E8.1. Comment on your findings.

9

Multi-Criteria Decision-Making – Outranking-Type Methods

9.1 Overview

Multi-criteria decision-making (MCDM) provides a robust framework for identifying the most suitable alternative(s) from a set defined by an alternatives-criteria matrix (ACM) (Nabavi et al., 2024). This chapter focuses on a specific class of MCDM techniques known as outranking-type methods, which evaluate alternatives through pairwise comparisons across multiple criteria. These methods establish an outranking relationship by determining the degree to which one alternative is preferred over another.

This chapter explores two prominent families of outranking-type MCDM methods, namely, ELECTRE (translated to: "Elimination and Choice Expressing the Reality") and PROMETHEE. The most popular variant from each of these families is detailed and illustrated using a simple numerical example to provide an intuitive understanding of their underlying mechanisms. The following sections in this chapter present these methods. Readers need not study them sequentially and can read either of them.

- Section 9.2 Elimination et Choix Traduisant la Realité (ELECTRE) Methods
- Section 9.3 Preference Ranking Organization Method for Enrichment Evaluations (PROMETHEE) Methods
- Section 9.4 Summary

Our MS Excel program, EMCDM, described in Chapter 10, includes two variants of ELECTRE.

The learning outcomes of this chapter are to (1) describe the principles and algorithms of outranking-type MCDM methods covered, namely, ELCTRE and PROMETHEE and (2) apply any of these methods to the ACM of an application.

DOI: 10.1201/9781003530435-9

TABLE 9.1

ACM Employed for Illustration of ELECTRE and PROMETHEE Methods (C1, C2 and C3 for Maximization; C4 for Minimization).

Alternatives	C1 (Max)	C2 (Max)	C3 (Max)	C4 (Min)
A1	0.78	117	6.2	35
A2	0.65	150	5.8	40
A3	0.83	110	4.9	30
Weight	0.30	0.25	0.20	0.25

Unless otherwise stated, the notation used throughout this chapter is as follows:

- $i \in \{1,2,\ldots,m\}$ and $j \in \{1,2,\ldots,n\}$, where m is the number of rows (i.e., alternatives) and n is the number of columns (i.e., criteria) in the ACM.
- f_{ij} is the value of the jth criterion for the ith alternative in the original ACM.
- F_{ij} is the value of f_{ij} after normalization/transformation.
- w_j is the weight assigned to the jth criterion and $\sum_{j=1}^{n} w_j = 1$.

Additionally, the ACM presented in Table 9.1 is employed to walk through the steps of the ELECTRE and PROMETHEE methods covered in this chapter. This ACM is made up of three alternatives (i.e., $m = 3$ and $i \in \{1,2,3\}$) and four criteria (i.e., $n = 4$ and $j \in \{1,2,3,4\}$). Out of the four criteria (C1, C2, C3 and C4), C1, C2 and C3 are benefit criteria to be maximized, whereas C4 is a cost criterion to be minimized. For illustration, the weights for the criteria are assumed to be $w_1 = 0.30$, $w_2 = 0.25$, $w_3 = 0.20$ and $w_4 = 0.25$. All three alternatives (A1, A2 and A3) are non-dominated, which means that, among them, the improvement in one criterion will inevitably result in degradation in at least one other criterion (Wang et al., 2024).

9.2 Elimination et Choix Traduisant la Realité (ELECTRE) Methods

The ELECTRE family of methods, originally proposed by Roy (1968), is a popular group of outranking techniques in MCDM, especially when dealing with complex problems involving conflicting criteria. These methods are designed to compare alternatives by constructing outranking relations through a series of pairwise comparisons across all criteria (Wang and Rangaiah, 2017), rather than relying on reference-based or aggregation-based scoring strategies.

The ELECTRE family has been developed and refined over several decades, with several variants such as ELECTRE I, II, III, IV, IS and TRI (Figueira et al., 2010; Govindan and Jepsen, 2016). It started with the introduction of ELECTRE I (Roy, 1968). Then, each subsequent variant addresses specific limitations or adapts to different decision-making contexts. ELECTRE I is the foundational model, developed for binary outranking decisions in choice problems, but it lacks the ability to provide rankings among alternatives (Govindan and Jepsen, 2016). Its improved version, ELECTRE II (Roy and Bertier, 1973), includes a mechanism for complete and partial ranking, making it more suitable for decision scenarios requiring prioritization (Lima and Soares, 2011).

Pseudo-criteria and credibility indices are introduced in ELECTRE III (Roy, 1978) to enable the handling of more complicated information and allow for more nuanced preference modeling in complex real-world applications (Figueira et al., 2010). ELECTRE IV (Roy and Hugonnard, 1982) extends these capabilities to settings with no available thresholds or preferences, making it applicable when limited quantitative information is present (Hokkanen and Salminen, 1997). ELECTRE IS (Roy and Bouyssou, 1993), an improved variant of ELECTRE I, was developed to simplify parameter elicitation (e.g., indifference thresholds) and improve the robustness of the conclusions (Park et al., 2005). ELECTRE TRI (Yu, 1992) introduces a sorting approach rather than ranking or choosing; it is designed to classify alternatives into predefined categories based on performance profiles, which makes it ideal for applications like credit scoring or risk classification (Mousseau and Slowinski, 1998).

Among these variants, ELECTRE III stands out as the most widely used and recognized one for ranking alternatives (Figueira et al., 2010), which introduces the indifference, preference and veto thresholds in its algorithm, allowing more nuanced comparisons between alternatives (Liao et al., 2020). The steps in the ELECTRE III method are outlined below.

Step 1. Transform all the original criteria to maximization type:

$$F_{ij} = \begin{cases} f_{ij} & \textit{if criterion } j \textit{ is for maximization} \\ -f_{ij} & \textit{if criterion } j \textit{ is for minimization} \end{cases} \tag{9.1}$$

This transformation ensures that a higher F_{ij} always indicates a better performance.

NUMERICAL CALCULATIONS (USING ACM FROM TABLE 9.1)

For instance, for $i = 1$ and $j = 4$, since C4 is a minimization criterion:

$$F_{14} = -f_{14} = -35$$

Similar transformations are performed for all other values in the ACM across the $m = 3$ alternatives and $n = 4$ criteria. The complete results, forming the transformed ACM, are shown in Table 9.2.

TABLE 9.2

Transformed ACM for ELECTRE III Walk-Through.

Alternatives	C1	C2	C3	C4
A1	0.78	117	6.2	–35
A2	0.65	150	5.8	–40
A3	0.83	110	4.9	–30

Step 2. Construct the concordance matrix by pairwise comparison; this matrix has m rows and m columns, corresponding to m alternatives. This construction is as follows:

Let a and b represent any two alternatives chosen for pairwise comparison. The indifference threshold (Q_j) defines the range of variations for the criterion j within which the decision-maker perceives no significant difference between the two alternatives. In other words, it represents the interval where the performance of two alternatives on a given criterion is considered indistinguishable. When the difference in performance between two alternatives for a specific criterion exceeds the preference threshold (P_j), a clear preference is assigned to the alternative with the better value (i.e., higher F_{ij}). The veto threshold (V_j) is used to disqualify an alternative relative to another when the difference in performance for the criterion j is too large to be deemed acceptable.

The Q_j, P_j and V_j (all non-negative) can be given by decision-makers based on the context of the application. So, these inputs will be subjective. Rogers and Bruen (1998) stated that these thresholds can be determined as a proportion of the range of criterion values. In the numerical example presented here, as well as in our EMCDM program (described in Chapter 10), the thresholds: Q_j, P_j and V_j are, respectively, set to 10%, 20% and 80% of the range of the criterion j (Lee and Rangaiah, 2009). These can be adjusted as and when necessary.

Each element of the concordance matrix, that is, $C(a,b)$, is computed using:

$$C(a,b) = \sum_{j=1}^{n} w_j C_j(a,b), \quad a,b \in \{1,2,\ldots,m\} \tag{9.2a}$$

$$\text{Here, } C_j(a,b) = \begin{cases} 1 & \text{if } F_j(b) - F_j(a) \le Q_j \\ 0 & \text{if } F_j(b) - F_j(a) > P_j \\ \dfrac{P_j - [F_j(b) - F_j(a)]}{P_j - Q_j} & \text{if } Q_j < F_j(b) - F_j(a) \le P_j \end{cases} \tag{9.2b}$$

Essentially, $C_j(a,b)$ represents the degree of support from the criterion j for the assertion that alternative a is at least as good as alternative b under this criterion. Here, $0 \le C_j(a,b) \le 1$. A value of $C_j(a,b) = 1$ means that criterion j fully supports this assertion, indicating that a performs as good as or better than b. Conversely, $C_j(a,b) = 0$ signifies that the performance gap exceeds the preference threshold P_j, meaning that a clearly is worse than b on criterion j. For the other $C_j(a,b)$ values between 0 and 1, the difference lies between Q_j and P_j, representing a partial support: Values closer to 1 suggest near indifference between a and b, while values closer to 0 imply that a is much less favorable than b.

A special case arises when an alternative is compared with itself (i.e., $a = b$); in this situation, $C_j(a,b) = 1$ according to Eq. (9.2b). Next, $C(a,b)$ aggregates these degrees of support across all criteria, multiplied by their weights, to express the overall strength of evidence that a is at least as good as b.

NUMERICAL CALCULATIONS

As stated above, the thresholds: Q_j, P_j and V_j for each criterion j are assumed to be, respectively, 10%, 20% and 80% of the range of the criterion. For instance, for $j = 2$, the thresholds of criterion C2 (whose maximum and minimum are, respectively, 150 and 110) are calculated as follows:

$$Q_2 = 0.1 \times (150 - 110) = 4$$

$$P_2 = 0.2 \times (150 - 110) = 8$$

$$V_2 = 0.8 \times (150 - 110) = 32$$

Comparing the two alternatives of $a=3$ and $b=1$ with respect to C2, $F_2(1)-F_2(3)=117-110=$ 7. Since this difference is between $Q_2=4$ and $P_2=8$,

$$C_2(3,1)=\frac{8-[117-110]}{8-4}=0.25$$

It should be noted that $C_j(a,b)$ need not be the same as $C_j(b,a)$ in the concordance matrix, according to Eq. (9.2b). For example, $C_2(1,3)=1$ as $F_2(3)-F_2(1)=110-117=-7$, which is less than $Q_2=4$.

Similarly, pairwise comparisons between these two alternatives are performed across the remaining three criteria; the computed values are: $C_1(3,1)=1$, $C_3(3,1)=0$ and $C_4(3,1)=1$. This is followed by the computation of the weighted aggregate, $C(3,1)=\sum_{j=1}^{n} w_j C_j(3,1)=0.6125$ using the assigned weights in Table 9.1. The complete concordance matrix for all three alternatives considering all four criteria is displayed in Table 9.3. Note that, according to Eq. (9.2b), all diagonal elements are equal to 1 (representing an alternative compared with itself).

TABLE 9.3

Concordance Matrix for ELECTRE III Walk-Through.

$C(a,b)$	A1	A2	A3
A1	1	0.75	0.45
A2	0.25	1	0.45
A3	0.6125	0.55	1

Step 3. Construct the discordance matrix under each criterion, that is, $D_j(a,b)$.

$$D_j(a,b)=\begin{cases} 1 & \text{if } F_j(b)-F_j(a)>V_j \\ 0 & \text{if } F_j(b)-F_j(a)\le P_j \\ \dfrac{F_j(b)-F_j(a)-P_j}{V_j-P_j} & \text{if } P_j<F_j(b)-F_j(a)\le V_j \end{cases} \tag{9.3}$$

Here, $0 \le D_j(a,b) \le 1$. As the counterpart of $C_j(a,b)$, $D_j(a,b)$ reflects the degree to which, for a criterion j, the performance of alternative a is significantly worse than that of alternative b. $D_j(a,b)=1$ means the difference surpasses the veto threshold V_j, signaling strong evidence from criterion j that a is significantly worse than b on this criterion. At the other extreme, $D_j(a,b)=0$ means criterion j provides no indication that a is significantly worse than b.

When an alternative is compared to itself (i.e., $a=b$), then $D_j(a,b)=0$. There will be n discordance matrices, each corresponding to one of n criteria.

NUMERICAL CALCULATIONS

For instance, for $j=2$, comparing the two alternatives of $a=3$ and $b=1$: Since $F_j(b)-F_j(a)=117-110=7 \le P_j=8$ in this case, we get $D_2(3,1)=0$. Likewise, pairwise comparisons of the other alternatives under the second criterion are conducted and the resulting discordance matrix, $D_2(a,b)$, is presented in Table 9.4. Also, this table shows discordance matrices for the remaining three criteria, derived in the same manner.

TABLE 9.4

Discordance Matrices for ELECTRE III Walk-Through.

$D_1(a,b)$	A1	A2	A3
A1	0	0	0.1296
A2	0.8704	0	1
A3	0	0	0
$D_2(a,b)$	**A1**	**A2**	**A3**
A1	0	1	0
A2	0	0	0
A3	0	1	0
$D_3(a,b)$	**A1**	**A2**	**A3**
A1	0	0	0
A2	0.1795	0	0
A3	1	0.8205	0
$D_4(a,b)$	**A1**	**A2**	**A3**
A1	0	0	0.5
A2	0.5	0	1
A3	0	0	0

Step 4. Construct the credibility matrix $S(a,b)$ of size $m \times m$ as follows:

$$S(a,b) = \begin{cases} C(a,b) & \text{if } D_j(a,b) \le C(a,b) \forall j \\ C(a,b) \prod_{j \in J} \frac{1 - D_j(a,b)}{1 - C(a,b)} & \text{otherwise} \end{cases} \tag{9.4}$$

Here, for the product term, the set, $J = \{ j | D_j(a,b) \rangle C(a,b) \}$. The quantity, $S(a,b)$, expresses the overall credibility or strength of the assertion that alternative a outranks alternative b considering both the concordance and discordance assessments.

NUMERICAL CALCULATIONS

For instance, consider the two alternatives: $a = 3$ and $b = 1$. The condition of $D_j(a,b) \le C(a,b) \forall j$ is not satisfied because $D_3(3,1) = 1$ (Table 9.4) and $C(3,1) = 0.6125$ (Table 9.3). Consequently, $J = \{3\}$ and so

$$S(3,1) = C(3,1) \prod_{j \in J} \frac{1 - D_j(3,1)}{1 - C(3,1)} = 0.6125 \left(\frac{1-1}{1-0.6125} \right) = 0$$

Consider two other alternatives: $a = 3$ and $b = 3$. The condition of $D_j(a,b) \le C(a,b) \forall j$ is satisfied because $D_1(3,3) = D_2(3,3) = D_3(3,3) = D_4(3,3) = 0.0$ (Table 9.4) and $C(3,3) = 1.0$ (Table 9.3). Consequently, $S(3,3) = C(3,3)$ = 1.0. Similarly, the remaining elements of the credibility matrix are computed. The complete credibility matrix is presented in Table 9.5.

TABLE 9.5

Credibility Matrix for ELECTRE III Walk-Through.

$S(a,b)$	A1	A2	A3
A1	1	0	0.4091
A2	0.0288	1	0
A3	0	0	1

Step 5. Finally, calculate the outranking score (R_a) for each alternative a.

$$R_a = \sum_{b=1}^{m} S(a,b) - \sum_{b=1}^{m} S(b,a) \tag{9.5}$$

Here, the term $\sum_{b=1}^{m} S(a,b)$ represents the strength of alternative a in outranking the other alternatives (including itself). Conversely, the term $\sum_{b=1}^{m} S(b,a)$ reflects the extent to which alternative a is outranked by the others (including itself). Hence, R_a reflects the overall strength of alternative a in these outranking relations. A larger R_a indicates that the alternative a is strongly preferred in pairwise comparisons and hence more desirable. Besides, R_a can be negative, which means that the alternative a is more frequently outranked by others than it outranks them, indicating a relatively weaker overall standing in the decision context. The alternative with the largest outranking score, R_a, is top-ranked and recommended to the decision-maker.

NUMERICAL CALCULATIONS

For instance, the outranking score for alternative, $a = 1$, using $S(a,b)$ values in Table 9.5, is:

$$R_1 = \sum_{b=1}^{m} S(1,b) - \sum_{b=1}^{m} S(b,1) = (1 + 0 + 0.4091) - (1 + 0.0288 + 0) = 0.3803$$

Similarly, the outranking scores of all alternatives are calculated as $R_1 = 0.3803$, $R_2 = 0.0288$ and $R_3 = -0.4091$. Accordingly, the ranking is A1 ≻ A2 ≻ A3, with A1 being the top-ranked alternative by ELECTRE III.

The advantages of the ELECTRE family are as follows: (1) By relying on outranking relations instead of additive aggregation, ELECTRE methods can express non-compensatory preferences; so, a very poor score for one criterion cannot be fully offset by excellence on others. (2) The method naturally accommodates qualitative, quantitative and partially ordered data because concordance–discordance checks are based only on pairwise comparisons rather than explicit utility scales. (3) Several well-defined ELECTRE variants (I, II, III, IV, IS, V, TRI, etc.) can be picked and used by decision-makers, following a familiar conceptual framework. (4) No normalization of the original ACM is required because the method relies on criterion-specific thresholds (indifference, preference and veto), which inherently account for differences in measurement scales.

The limitations of the ELECTRE family are as follows: (1) Besides weights, most variants require additional inputs such as preference thresholds for every criterion, which can be cognitively demanding to elicit and may increase subjectivity. (2) The use of veto and discordance in ELECTRE methods can result in some alternatives being incomparable (i.e., no clear preference can be established between them due to conflicting evidence such as strong concordance support overridden by high discordance or veto). While this reflects real-world uncertainty and can be useful, too many such instances of incomparability may frustrate decision-makers, who typically prefer a complete ranking of alternatives. (3) With an increasing number of alternatives and/or criteria, the comparison matrices (i.e., concordance, discordance and credibility) grow quickly, making ELECTRE methods computationally intensive.

9.3 Preference Ranking Organization Method for Enrichment Evaluations (PROMETHEE) Methods

The PROMETHEE family, first put forward by Brans et al. (1982), belongs to the same class of outranking type as ELECTRE methods. It relies on pairwise comparisons of alternatives on every criterion, converting each performance difference into a preference degree through one of six generalized preference functions outlined in Brans et al. (1986) and then aggregating those degrees (usually with simple additive weights) into multi-criteria preference indices. From these indices, it then computes the leaving and entering flows that express how much each alternative outranks or is outranked by the others.

Like the ELECTRE family, the PROMETHEE family also has several variants, such as PROMETHEE I, II, III, IV, V and VI (Behzadian et al., 2010). PROMETHEE I (Brans et al., 1982) provides a partial ranking of alternatives based on preference flows, allowing for incomparability between some options. PROMETHEE II (Brans et al., 1986) extends this by delivering a complete ranking using net preference flows, making all alternatives directly comparable (Alinezhad and Khalili, 2019). PROMETHEE III introduces interval evaluations to incorporate uncertainty, considering alternatives as indifferent if their preference flows are very close (Brans and De Smet, 2005; Fernández-Castro and Jiménez, 2005).

The next variant, PROMETHEE IV, is adapted for problems involving a continuous set of alternatives, often using integral calculus to evaluate preference distributions (Brans and De Smet, 2005; Albuquerque and Montenegro, 2016). PROMETHEE V (Brans and Mareschal, 1992) incorporates optimization under constraints by integrating integer LP, which is useful for selecting the best subset of actions under limited resources (Fernández-Castro and Jiménez, 2005). PROMETHEE VI (Brans and Mareschal, 1995) handles imprecise criteria weights by defining them as intervals instead of fixed values; it calculates the stability range of possible net flows for each alternative, visualizing the results in a profile called "human brain" (Figueira et al., 2005).

Among these variants, PROMETHEE II is consistently reported as the most widely applied method of the family (Torbacki, 2021; Taherdoost and

Madanchian, 2023) because it is computationally straightforward and conceptually simple and it is the first method in the PROMETHEE family, which can generate a complete ranking of alternatives. Hence, the steps of PROMETHEE II are outlined below.

Step 1. Choose a generalized preference function $P_j(d)$ from Table 9.6 for every criterion j in the ACM. The preference function need not be the same for all criteria. In this table, d represents the difference between the two alternatives (a and b) with respect to the criterion j. In this step, the indifference threshold q (non-negative) and/or the preference threshold p (non-negative) must be specified by the decision-maker. The indifference threshold q sets the largest gap on a criterion that the decision-maker still regards as negligible (i.e., when the difference in the values of a criterion for two alternatives is less than or equal to q, the two alternatives are treated as essentially equal). The preference threshold p marks the point where the gap becomes clearly significant, giving full or strict preference to the better performer.

TABLE 9.6

Generalized Preference Functions Used in PROMETHEE Methods (Brans et al., 1986).

Generalized Preference Function	Definition	Parameters
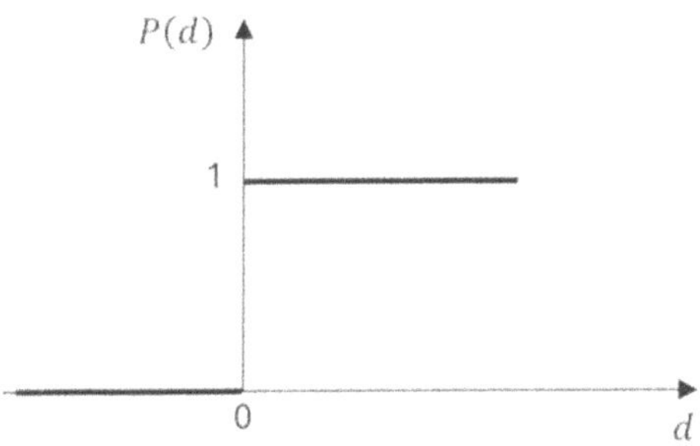	Type I: Usual criterion $P(d)=\begin{cases}0 & \text{if } d\le 0\\1 & \text{if } d>0\end{cases}$	Nil
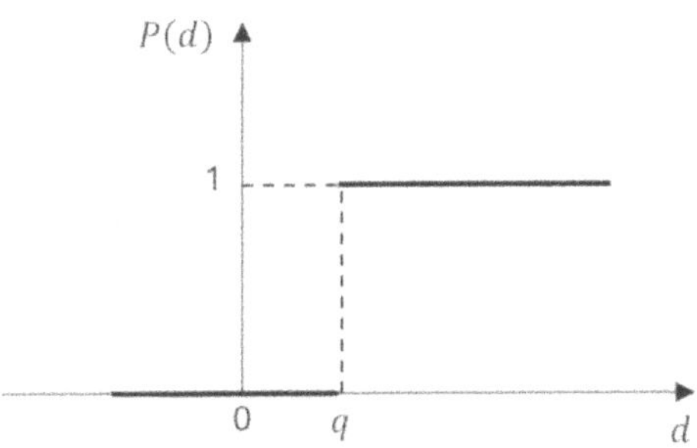	Type II: Quasi-criterion $P(d)=\begin{cases}0 & \text{if } d\le q\\1 & \text{if } d>q\end{cases}$	q

(Continued)

TABLE 9.6 ***(Continued)***

Generalized Preference Functions Used in PROMETHEE Methods (Brans et al., 1986).

Generalized Preference Function	Definition	Parameters
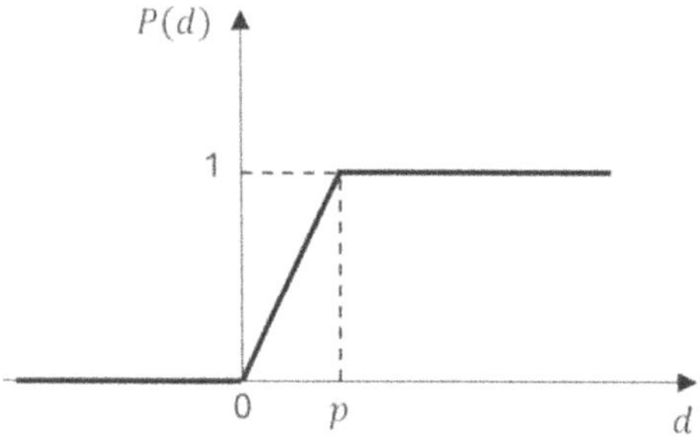	Type III: Criterion with linear preference $P(d)=\begin{cases}0 & \text{if } d\le 0\\ \dfrac{d}{p} & \text{if} 0<d\le p\\ 1 & \text{if } d>p\end{cases}$	p
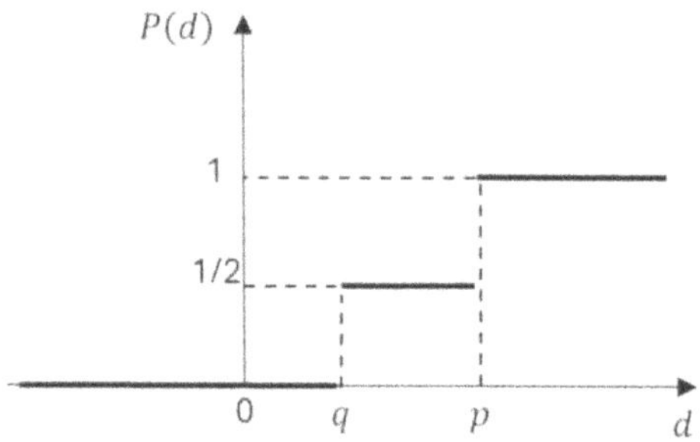	Type IV: Level criterion $P(d)=\begin{cases}0 & \text{if } d\le q\\ \dfrac{1}{2} & \text{if } q<d\le p\\ 1 & \text{if } d>p\end{cases}$	q,p
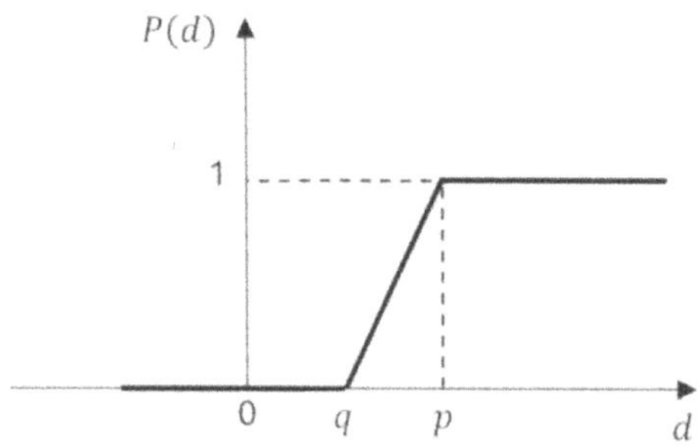	Type V: Criterion with linear preference and indifference area $P(d)=\begin{cases}0 & \text{if } d\le q\\ \dfrac{d-q}{p-q} & \text{if } q<d\le p\\ 1 & \text{if } d>p\end{cases}$	q,p
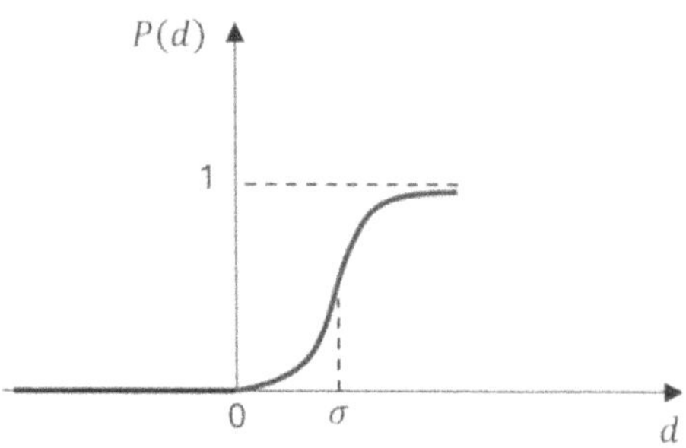	Type VI: Gaussian criterion $P(d)=\begin{cases}0 & \text{if } d\le 0\\ 1-e^{\frac{-d^2}{2\sigma^2}} & \text{if } d>0\end{cases}$	σ (standard deviation)

The general guidance on choosing a preference function for a criterion is as follows: Type I is used when any nonzero difference implies a strict preference, making it suitable for criteria where even the slightest improvement is meaningful. Type II is appropriate when only differences beyond a certain threshold q are considered significant, meaning that there is an indifference zone. Type III fits situations where preference increases linearly with the difference from zero up to a predefined threshold p, ideal for criteria with proportional sensitivity. Type IV introduces three zones: indifference up to q, a middle range of weak preference between q and p and strict preference beyond p, making it useful when gradations of importance are recognized. Type V further refines Type IV by assuming small differences up to q are negligible, but preferences increase linearly between q and p. Lastly, Type VI is best suited for continuous, smooth data where the preference increases gradually, commonly applied in criteria requiring sensitivity to nuanced differences.

NUMERICAL CALCULATIONS (USING ACM FROM TABLE 9.1)

For instance, we have selected the preference functions for the four criteria and specified the corresponding parameter values, as presented in Table 9.7. These selections and values are for demonstration and they can be changed.

TABLE 9.7

ACM with Preference Functions and Parameter Values for the PROMETHEE II Walk-Through.

Alternatives	**C1**	**C2**	**C3**	**C4**
A1	0.78	117	6.2	– 35
A2	0.65	150	5.8	– 40
A3	0.83	110	4.9	– 30
Preference Function	**Type I**	**Type V**	**Type III**	**Type II**
q	–	5	–	8
p	–	10	0.8	–

Step 2. Compute preference degree for each pair of alternatives (a and b) under criterion j.

$$d_{ab}^{(j)} = \begin{cases} f_j(a) - f_j(b) & \text{if criterion } j \text{ is to maximize} \\ f_j(b) - f_j(a) & \text{if criterion } j \text{ is to minimize} \end{cases} \tag{9.6a}$$

As such, positive $d_{ab}^{(j)}$ always favors alternative a. Next, $d_{ab}^{(j)}$ is fed into the chosen $P_j(d)$ function to compute the preference degree of alternatives a and b under criterion j as follows:

$$\text{Preference degree} = P_j\left(d_{ab}^{(j)}\right) \tag{9.6b}$$

Overall, $P_j\left(d_{ab}^{(j)}\right) = 0$ signifies an indifference between a and b under criterion j. In contrast, $P_j\left(d_{ab}^{(j)}\right) = 1$ means strict preference of a over b with respect to the criterion. A special case is when an alternative is compared to itself (i.e., $a = b$), it always holds that $P_j\left(d_{ab}^{(j)}\right) = 0$.

NUMERICAL CALCULATIONS

For instance, consider the criterion $j = 3$ (maximization/benefit criterion) and compare the two alternatives of $a = 1$ and $b = 2$. The performance gap is:

$$d_{12}^{(3)} = f_3(1) - f_3(2) = 6.2 - 5.8 = 0.4$$

As given in Table 9.7, this criterion employs Type III with a preference threshold $p = 0.8$. Because $0 < d_{12}^{(3)} \leq p$, using the Type III preference definition from Table 9.6, the preference degree is obtained:

$$P_3\left(d_{12}^{(3)}\right) = \frac{d_{12}^{(3)}}{p} = \frac{0.4}{0.8} = 0.5$$

Thus, alternative $a = 1$ is accorded with a moderate preference of 0.5 over alternative $b = 2$ on criterion 3.

Now, for the same benefit criterion $j = 3$, let us compare the two alternatives the other way around, that is, $a = 2$ and $b = 1$. The performance gap is:

$$d_{21}^{(3)} = f_3(2) - f_3(1) = 5.8 - 6.2 = -0.4$$

This criterion employs Type III and $d_{21}^{(3)} \leq 0$. Using the Type III preference definition from Table 9.6, the preference degree is obtained:

$$P_3\left(d_{21}^{(3)}\right) = 0$$

Similarly, following this comparison and calculation using the chosen preference function, the preference degree for other pairs of alternatives under each criterion can be obtained. The complete preference degree matrices are presented in Table 9.8. As observed, each preference degree matrix is not symmetric along the principal diagonal, since $P_j\left(d_{ab}^{(j)}\right)$ does not necessarily equal $P_j\left(d_{ba}^{(j)}\right)$.

TABLE 9.8

Preference Degree Matrices for PROMETHEE II Walk-Through.

$P_1(d)$	**A1**	**A2**	**A3**
A1	0	1	0
A2	0	0	0
A3	1	1	0
$P_2(d)$	**A1**	**A2**	**A3**
A1	0	0	0.4
A2	1	0	1
A3	0	0	0
$P_3(d)$	**A1**	**A2**	**A3**
A1	0	0.5	1
A2	0	0	1
A3	0	0	0
$P_4(d)$	**A1**	**A2**	**A3**
A1	0	0	0
A2	0	0	0
A3	0	1	0

Step 3. Aggregate the preference degrees to build the multi-criteria preference matrix $\pi(a,b)$.

$$\pi(a,b) = \sum_{j=1}^{n} w_j \, P_j\left(d_{ab}^{(j)}\right) \tag{9.7}$$

NUMERICAL CALCULATIONS

For instance, for the two alternatives of $a = 1$ and $b = 2$:

$$\pi(1,2) = \sum_{j=1}^{4} w_j \, P_j\left(d_{12}^{(j)}\right) = 0.30 \times 1 + 0.25 \times 0 + 0.20 \times 0.5 + 0.25 \times 0 = 0.4$$

Similarly, the remaining elements of the multi-criteria preference matrix can be computed. The complete matrix is presented in Table 9.9. This matrix is not symmetric, as the individual preference matrices are not symmetric.

TABLE 9.9

Multi-Criteria Preference Matrix for PROMETHEE II Walk-Through.

Alternatives	A1	A2	A3
A1	0	0.4	0.3
A2	0.25	0	0.45
A3	0.3	0.55	0

Step 4. Calculate the outranking score R_a for each alternative a.

$$R_a = \sum_{b=1}^{m} \pi(a,b) - \sum_{b=1}^{m} \pi(b,a) \tag{9.8}$$

Here, the sum, $\sum_{b=1}^{m} \pi(a,b)$, is the positive flow, measuring the extent to which alternative a outranks all other alternatives (including itself). Conversely, the sum, $\sum_{b=1}^{m} \pi(b,a)$, is the negative flow, indicating how strongly a is outranked by the other alternatives (including itself). Hence, R_a reflects the overall strength of alternative a in these outranking relations. The alternative with the largest R_a is top-ranked and recommended to the decision-maker.

NUMERICAL CALCULATIONS

For instance, for alternative $a = 1$:

$$R_1 = \sum_{b=1}^{m} \pi(1,b) - \sum_{b=1}^{m} \pi(b,1) = (0+0.4+0.3) - (0+0.25+0.3) = 0.15$$

Analogously, outranking scores of all alternatives are calculated as $R_1 = 0.15$, $R_2 = -0.25$ and $R_3 = 0.1$. Accordingly, the ranking is A1 > A3 > A2, with A1 being the top-ranked alternative by PROMETHEE II.

The advantages of the PROMETHEE family are as follows: (1) PROMETHEE methods deliver a complete ranking for every set of alternatives, sparing decision-makers the frustration of incomparable and possibly many "ties" as in ELECTRE methods. (2) A coherent suite of variants uses the same underlying preference mechanism and so decision-makers can switch from one variant of PROMETHEE to another with ease. (3) No normalization of the original ACM is required because the use of preference functions and associated thresholds allows each criterion to be evaluated directly in its own units. (4) PROMETHEE methods are computationally slightly lighter than ELECTRE methods, as their steps are more straightforward and involve fewer calculations.

The limitations of the PROMETHEE family are as follows: (1) Beyond assigning criterion weights, decision-makers must also choose a preference function for each criterion and specify parameters such as the indifference threshold q, preference threshold p, or Gaussian spread σ; eliciting these inputs can be cognitively demanding and inject subjectivity. (2) Because PROMETHEE requires a pairwise comparison for every pair of alternatives, the size of the preference matrix grows rapidly, making it computationally intensive.

9.4 Summary

Main points in this chapter on outranking-type MCDM methods are as follows:

1. This chapter describes two prominent outranking-type MCDM families (ELECTRE and PROMETHEE). These methods evaluate alternatives using pairwise comparisons to establish outranking relationships.
2. The most widely used variant of each family, namely, ELECTRE III and PROMETHEE II, is described and illustrated using a simple numerical example, enabling intuitive understanding of their respective mechanisms.
3. Outranking-type MCDM methods assess the extent to which one alternative outranks another, incorporating notions such as indifference and preference thresholds. These thresholds allow decision-makers to express nuanced preferences that might not be captured by purely additive and multiplicative approaches.
4. Each method in this chapter is systematically presented with a clear explanation of its theoretical foundation, algorithmic steps and numerical calculations using a common ACM. This helps readers grasp the practical implementation of each method.
5. The chapter discusses the advantages and limitations associated with the ELECTRE and PROMETHEE families. Both methods do not require normalization of ACM. However, they require additional parameters beyond criterion weights, such as threshold values and/or preference functions, which enhance flexibility but also introduce subjectivity and cognitive burden for decision-makers.

6. Finally, using the common ACM presented in Table 9.1, both ELECTRE III and PROMETHEE II identify A1 as the top-ranked alternative. The ranking order of the remaining alternatives differs between the two methods, reflecting their underlying methodological distinctions. These ranking results cannot be generalized and it is advisable to apply multiple MCDM methods when solving real-world problems.

References

Albuquerque, P. H. M., & Montenegro, M. R. (2016). PROMETHEE IV through kernel density estimation. *Communications in Statistics-Theory and Methods, 45*(18), 5355–5362.

Alinezhad, A., & Khalili, J. (2019). PROMETHEE I-II-III methods. In *New methods and applications in Multiple Attribute Decision Making (MADM)* (Vol. 277). Springer.

Behzadian, M., Kazemzadeh, R. B., Albadvi, A., & Aghdasi, M. (2010). PROMETHEE: A comprehensive literature review on methodologies and applications. *European Journal of Operational Research, 200*(1), 198–215.

Brans, J. P., & De Smet, Y. (2005). PROMETHEE methods. In *Multiple criteria decision analysis: State of the art surveys* (pp. 187–219). New York, NY: Springer New York.

Brans, J. P., & Mareschal, B. (1992). PROMETHEE V: MCDM problems with segmentation constraints. *INFOR: Information Systems and Operational Research, 30*(2), 85–96.

Brans, J. P., & Mareschal, B. (1995). The PROMETHEE VI procedure: How to differentiate hard from soft multicriteria problems. *Journal of Decision Systems, 4*(3), 213–223.

Brans, J. P., Nadeau, R., & Landry, M. (1982). L'ingénierie de la décision. Elaboration d'instruments d'aide à la décision. La méthode PROMETHEE. In *L'Aide à la Décision: Nature, Instruments et Perspectives d'Avenir* (pp. 183–213). Québec: Presses de l'Université Laval..

Brans, J. P., Vincke, P., & Mareschal, B. (1986). How to select and how to rank projects: The PROMETHEE method. *European Journal of Operational Research, 24*(2), 228–238.

Fernández-Castro, A. S., & Jiménez, M. (2005). PROMETHEE: An extension through fuzzy mathematical programming. *Journal of the Operational Research Society, 56*(1), 119–122.

Figueira, J. R., Greco, S., Ehrogott, M., Brans, J. P., & Mareschal, B. (2005). PROMETHEE methods. In *Multiple criteria decision analysis: State of the art surveys* (pp. 163–186). New York, NY: Springer.

Figueira, J. R., Greco, S., Roy, B., & Słowiński, R. (2010). ELECTRE methods: Main features and recent developments. In *Handbook of multicriteria analysis* (pp. 51–89). Berlin, Heidelberg: Springer.

Govindan, K., & Jepsen, M. B. (2016). ELECTRE: A comprehensive literature review on methodologies and applications. *European Journal of Operational Research, 250*(1), 1–29.

Hokkanen, J., & Salminen, P. (1997). ELECTRE III and IV decision aids in an environmental problem. *Journal of Multi-Criteria Decision Analysis, 6*(4), 215–226.

Lee, E. S. Q., & Rangaiah, G. P. (2009). Optimization of recovery processes for multiple economic and environmental objectives. *Industrial & Engineering Chemistry Research*, *48*(16), 7662–7681.

Liao, H., Wu, X., Mi, X., & Herrera, F. (2020). An integrated method for cognitive complex multiple experts multiple criteria decision making based on ELECTRE III with weighted Borda rule. *Omega*, *93*, 102052.

Lima, A., & Salazar Soares, V. (2011). Multi criteria decision making models: An overview on electre methods. *Universidade Portucalense, Centro de Investigação em Gestão e Economia (CIGE) Working Papers*, (21/2011).

Mousseau, V., & Slowinski, R. (1998). Inferring an ELECTRE TRI model from assignment examples. *Journal of Global Optimization*, *12*, 157–174.

Nabavi, S. R., Wang, Z., & Rangaiah, G. P. (2024). Sensitivity assessment of multi-criteria decision-making methods in chemical engineering optimization applications. In F. Gómez-Castro, & V. Rico-Ramírez (Eds.), *Optimization in chemical engineering: Deterministic, meta-heuristic and data-driven techniques* (pp. 283–304). Berlin, Boston: De Gruyter. https://doi.org/10.1515/9783111383439-008

Park, S., Kim, J., & Kim, S. K. (2005). A mixed-integer programming model to draw the concordance level and the kernel set for the implementation of ELECTRE IS. *Journal of Korean Institute of Industrial Engineers*, *31*(4), 265–276.

Rogers, M., & Bruen, M. (1998). Choosing realistic values of indifference, preference and veto thresholds for use with environmental criteria within ELECTRE. *European Journal of Operational Research*, *107*(3), 542–551.

Roy, B. (1968). Classement et choix en présence de points de vue multiples. *Revue française d'informatique et de recherche opérationnelle*, *2*(8), 57–75.

Roy, B. (1978). ELECTRE III: Un algorithme de classements fondé sur une représentation floue des préférences en présence de critères multiples. *Cahiers du Centre d'études de Recherche Opérationnelle*, *20*, 3–24.

Roy, B., & Bertier, P. (1973). La méthode Electre II: une application au media-planning. In *Operational Research: Or; Proceedings of the IFORS International Conference on Operational Research* (pp. 291–302). North Holland Publ. Co.

Roy, B., & Bouyssou, D. (1993). *Aide multicritère à la décision: méthodes et cas* (Vol. 695). Paris: Economica.

Roy, B., & Hugonnard, J. C. (1982). Ranking of suburban line extension projects on the Paris metro system by a multicriteria method. *Transportation Research Part A: General*, *16*(4), 301–312.

Taherdoost, H., & Madanchian, M. (2023). Using PROMETHEE method for multi-criteria decision making: Applications and procedures. *Iris Journal of Economics & Business Management*, *1*(1).

Torbacki, W. (2021). A hybrid MCDM model combining DANP and PROMETHEE II methods for the assessment of cybersecurity in Industry 4.0. *Sustainability*, *13*(16), 8833.

Wang, Z., Nabavi, S. R., & Rangaiah, G. P. (2024). Multi-criteria decision making in chemical and process engineering: Methods, progress and potential. *Processes*, *12*(11), 2532. https://doi.org/10.3390/pr12112532

Wang, Z., & Rangaiah, G. P. (2017). Application and analysis of methods for selecting an optimal solution from the Pareto-optimal front obtained by multiobjective optimization. *Industrial & Engineering Chemistry Research*, *56*(2), 560–574.

Yu, W. (1992). *ELECTRE TRI (aspects méthodologiques et manuel d'utilisation)*. Document-Université de Paris-Dauphine, LAMSADE.

Exercises

E9.1 You are given an ACM consisting of three alternatives (A1, A2 and A3) and four criteria (C1, C2, C3 and C4), with the following values:

Alternatives	C1	C2	C3	C4
A1	1.39	358	0.161	2.857
A2	1.55	448	0.172	2.5
A3	1.88	328	0.204	3.333

Criteria C1 and C2 are benefit criteria to be maximized, whereas C3 and C4 are cost criteria to be minimized. The weights for the criteria are $w_1 = 0.25$, $w_2 = 0.23$, $w_3 = 0.33$ and $w_4 = 0.19$. Apply both the outranking-type MCDM methods covered in this chapter to rank the alternatives. Identify the top-ranked alternative by each method and compare them.

E9.2 Briefly explain the role of the veto threshold (V_j) in ELECTRE III. Conduct sensitivity analysis by changing their values for the ACM in E9.1. Comment on your findings.

E9.3 Briefly explain the role of the parameters q and p in PROMETHEE II. Conduct sensitivity analysis by changing their values for the ACM in E9.1. Comment on your findings.

E9.4 You are given an ACM consisting of four alternatives (A1, A2, A3 and A4) and three criteria (C1, C2 and C3), with the following values:

Alternatives	C1	C2	C3
A1	52.9	25.3	0.346
A2	46.6	23.5	0.367
A3	59.8	28.3	0.332
A4	57.8	25.9	0.308

Criteria C1 and C2 are benefit criteria to be maximized, whereas C3 is a cost criterion to be minimized. The weights for the criteria are $w_1 = 0.35$, $w_2 = 0.39$ and $w_3 = 0.26$. Apply both the outranking-type MCDM methods covered in this chapter to rank the alternatives. Identify the top-ranked alternative by each method and compare them.

E9.5 Describe the similarities and differences between the ELECTRE III and PROMETHEE II methods.

10

Programs and Software for Multi-Criteria Decision-Making

10.1 Overview

As multi-criteria decision-making (MCDM) continues to evolve in both theory and practice, the need for accessible and reliable programs/software has grown significantly. This chapter describes some multi-purpose and popular MCDM programs, software and libraries, each designed to support practitioners and researchers in applying diverse MCDM methods through user-friendly interfaces or programmable environments. Specifically, it describes five versatile MCDM tools across different platforms; they are one program in Microsoft Excel, open-source programming libraries (two in Python and one in the R language) and one standalone desktop software. All these programs are available readily and free of charge.

Section 10.2 presents a Microsoft Excel-based MCDM (EMCDM) program that offers an intuitive approach to many weighting and ranking/MCDM methods within the spreadsheet environment. EMCDM is implemented using Excel VBA (Visual Basic for Applications), allowing for interactive buttons and inputs (in worksheets) and automated calculations. Its use requires basic familiarity with MS Excel and no programming skills.

Sections 10.3 and 10.4 cover two comprehensive Python libraries for MCDM, namely, pymcdm and pyDecision. Section 10.5 describes RMCDA, a comprehensive R package for MCDM methods. Section 10.6 introduces SuperDecisions, a desktop application well-suited for using the AHP method and its generalization, the ANP method. Section 10.7 outlines several other MCDM programs/software. Section 10.8 presents general guidelines for reliable and robust MCDM. These sections are followed by Summary, References and Exercises (as in the previous chapters).

The learning outcomes of this chapter are to (1) identify the features, capabilities and platforms of five multi-purpose MCDM tools, namely, EMCDM, pymcdm, pyDecision, RMCDA and SuperDecisions; (2) select appropriate MCDM software or libraries based on the decision context, user needs and

DOI: 10.1201/9781003530435-10

computational preferences; and (3) apply these tools to MCDM applications and obtain the results of interest.

10.2 Microsoft Excel-Based MCDM (EMCDM) Program

The EMCDM program was initially developed by Wang and Rangaiah (2017) as Version 1, featuring the implementation of ten ranking/MCDM methods using Microsoft Excel VBA. Excel was chosen as the implementation platform because of its popularity in both academia and industry. Since then, EMCDM has been openly shared with hundreds of interested researchers, practitioners and students worldwide. It was subsequently enhanced in Wang et al. (2020) as Version 2 to add a total of eight weighting methods (four subjective and four objective) and expand the set of MCDM methods to 14. Further developments in Wang et al. (2021) led to Version 3, further increasing the number of MCDM methods to 16 and improving both computational performance and plot generation efficiency through optimizing the underlying VBA code for the various weighting and ranking methods.

As shown in Figure 10.1, EMCDM Version 3 implements 16 MCDM methods, listed in alphabetical order: CODAS, COPRAS, ELECTRE II, ELECTRE III, FUCA, GRA, LP technique for multidimensional analysis of preference (LINMAP), MABAC, MEW, MOORA, net flow method (NFM), SAW, TOPSIS, VIKOR, PROBID and simpler PROBID (sPROBID). Many of these are thoroughly described and illustrated in Chapters 7–9.

The eight weighting methods implemented in EMCDM Version 3, in alphabetical sequence, are the AHP method, BWM, entropy method, CRITIC method, mean method, SD method, statistical variance method and SWARA method. All these are described and illustrated in Chapters 4 and 5.

Although further enhancements are planned to incorporate additional methods and features (without compromising the program's current ease of use and functionality), the main interface of the latest stable release, EMCDM Version 3, is illustrated in Figure 10.1. To use EMCDM, decision-makers simply follow the five steps displayed in Figure 10.1. The "Overview" worksheet provides a brief user guide. The "Results from MOO" worksheet is where users begin by entering their ACM or importing Pareto-optimal solutions after solving their MOO problem in Step 1. Then, in Step 2, users supply key input details, including the number of objectives (criteria), decision variables, constraints and solutions (alternatives). The fields for decision variables and constraints, if not applicable, should be set to zero.

Step 3 involves specifying the type of each criterion, where "Max" represents a benefit type (to be maximized) and "Min" denotes a cost type (to be minimized). Next, in Step 4, decision-makers enter criteria weights. These can be input manually (ensuring the total equals unity) or automatically computed using one of the eight available weighting methods (e.g., AHP or entropy), which are displayed on the right-hand side of the main interface (Figure 10.1). Finally, in Step 5, users can select a specific MCDM method (e.g., TOPSIS) from the left panel to execute it and view its results generated in a separate worksheet; alternatively, click the "Run All Methods" button to generate results for all 16 MCDM methods implemented in EMCDM.

Another feature is that, in the top-left corner of the interface, there is a toggle switch to enable or disable plot generation. For MCDM problems involving only several alternatives, the numerical results alone may be sufficient to clearly understand the performance scores and rankings, making plots unnecessary. However, when dealing with many alternatives (such as numerous Pareto-optimal solutions found by MOO of an application), enabling plot generation is recommended, as it helps decision-makers visualize trade-offs across different criteria/objectives.

If decision-makers have generated charts by switching on plot generation and running more than one method, they can click the "Show All Charts" button to view and compare the results from these methods in a single worksheet. Additionally, after reviewing and noting/saving the required results, it is recommended to use the "Delete All Results" button to safely clear all previously generated results from the workbook. The EMCDM file can then be used for another application or study.

Figure 10.2 illustrates a representative example of the results obtained from executing the TOPSIS method using the EMCDM program. The

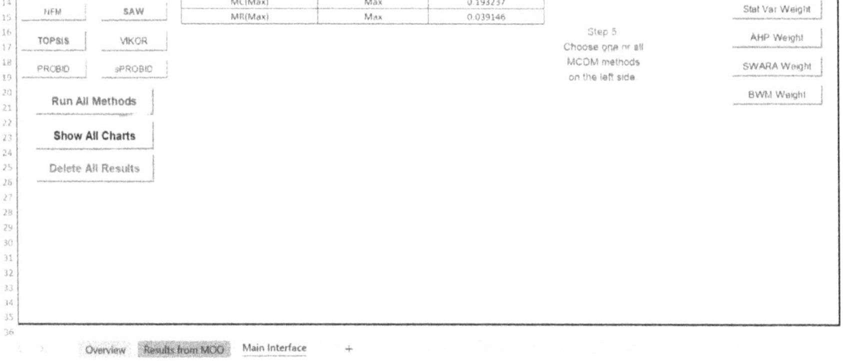

FIGURE 10.1
Main interface of the EMCDM program (version 3) with 16 MCDM methods and eight weighting methods.

analysis is based on an ACM (consisting of 100 alternatives or Pareto-optimal solutions for two criteria) generated by solving the bi-objective optimization problem from Exercise E2.4 in Chapter 2. In this problem, the objective (criterion) F1 is to be maximized, while F2 is to be minimized.

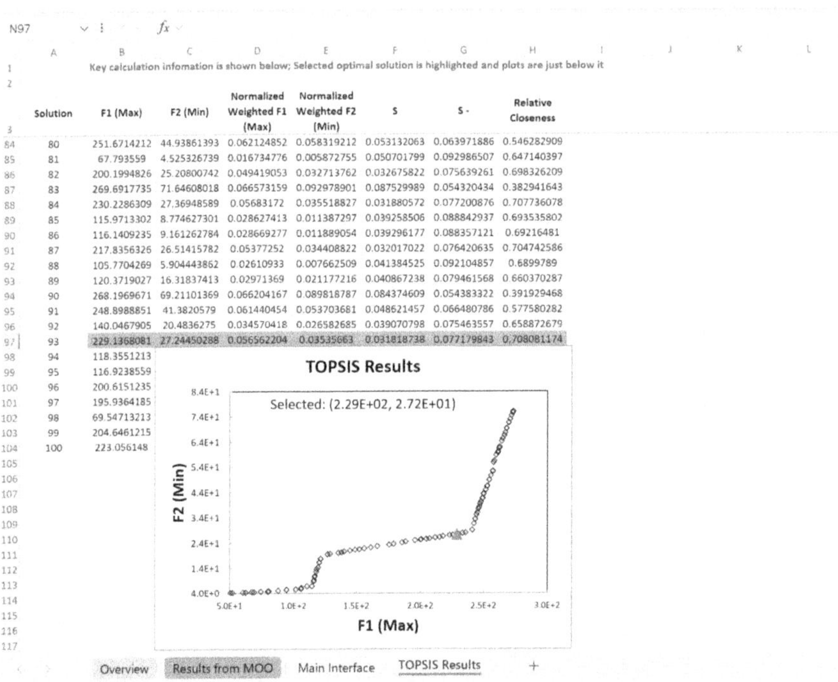

Solution	F1 (Max)	F2 (Min)	Normalized Weighted F1 (Max)	Normalized Weighted F2 (Min)	S	S -	Relative Closeness
80	251.6714212	44.93861393	0.062124852	0.058319212	0.053132063	0.063971886	0.546282909
81	67.793559	4.525326739	0.016734776	0.005872755	0.050701799	0.092986507	0.647140397
82	200.1994826	25.20800742	0.049419053	0.032713762	0.032675822	0.075639261	0.698326209
83	269.6917735	71.64608018	0.066573159	0.092978901	0.087529989	0.054320434	0.382941643
84	230.2286309	27.36948589	0.05683172	0.035518827	0.031880572	0.077200876	0.707736078
85	115.9713302	8.774627301	0.028627413	0.011387297	0.039258506	0.088842937	0.693535802
86	116.1409235	9.161262784	0.028669277	0.011889054	0.039296177	0.088357121	0.69216481
87	217.8356326	26.51415782	0.05377252	0.034408822	0.032017022	0.076420635	0.704742586
88	105.7704269	5.904443862	0.02610933	0.007662509	0.041384525	0.092104857	0.6899789
89	120.3719027	16.31837413	0.02971369	0.021177216	0.040867238	0.079461568	0.660370287
90	268.1969671	69.21101369	0.066204167	0.089818787	0.084374609	0.054383322	0.391929468
91	248.8988851	41.3820579	0.061440454	0.053703681	0.048621457	0.066480786	0.577580282
92	140.0467905	20.4836275	0.034570418	0.026582685	0.039070798	0.075463557	0.658872679
93	229.1368081	27.24450288	0.056562204	0.03535663	0.031818738	0.077179843	0.708081174
94	118.3551213						
95	116.9238559						
96	200.6151235						
97	195.9364185						
98	69.54713213						
99	204.6461215						
100	223.056148						

Super Decisions Main Window: AHP.sdmod: Unweighted Super Matrix

Cluster Node Labels		L1-Goal	L2-Criteria			L3-Alternatives			
		Goal_Node	C1	C2	C3	A1	A2	A3	A4
L1-Goal	Goal_Node	0.000000	0.000000	0.000000	0.000000	0.000000	0.000000	0.000000	0.000000
L2-Criteria	C1	0.626696	0.000000	0.000000	0.000000	0.000000	0.000000	0.000000	0.000000
	C2	0.279688	0.000000	0.000000	0.000000	0.000000	0.000000	0.000000	0.000000
	C3	0.093616	0.000000	0.000000	0.000000	0.000000	0.000000	0.000000	0.000000
L3-Alternatives	A1	0.000000	0.529640	0.267456	0.040746	0.000000	0.000000	0.000000	0.000000
	A2	0.000000	0.038790	0.583045	0.082529	0.000000	0.000000	0.000000	0.000000
	A3	0.000000	0.174103	0.108638	0.390851	0.000000	0.000000	0.000000	0.000000
	A4	0.000000	0.257467	0.040862	0.485875	0.000000	0.000000	0.000000	0.000000

Done

FIGURE 10.2
Example of TOPSIS results using the EMCDM program.

Here are the priorities.

Name	Normalized by Cluster
A1	0.41054
A2	0.19511
A3	0.17608
A4	0.21827

FIGURE 10.2 ***(Continued)***
Example of TOPSIS results using the EMCDM program.

Figure 10.2 presents the intermediate calculations, including normalized weighted values and the relative closeness scores, using equal weights of 0.5 for both F1 and F2. The optimal solution (alternative), ranked highest based on relative closeness, is highlighted in purple. A corresponding scatter plot is shown directly below the highlighted row, with the top-ranked solution marked by a red star for clear visualization.

Figure 10.3 shows the chosen alternatives by five MCDM methods (GRA, MABAC, PROBID, SAW and TOPSIS) for another application (machine tool selection in Chakraborty, 2011) having 21 alternatives, each assessed by seven criteria (i.e., ACM of 21 rows and seven columns). These results are presented in different formats: table, bar chart and radar diagram, for users' convenience. For this application, GRA, MABAC and PROBID recommended the same alternative, whereas SAW and TOPSIS suggested different alternatives.

Readers interested in the EMCDM program may request a free copy of it by contacting the authors via email: Dr. Wang (zywang@suss.edu.sg) or Prof. Rangaiah (chegpr@nus.edu.sg).

10.3 pymcdm Python Library

In recent years, Python has emerged as a convenient platform for MCDM applications due to its simplicity, readability and vast ecosystem of scientific libraries. One notable contribution to the MCDM community is pymcdm, a comprehensive and open-source Python library developed by Kizielewicz et al. (2023). Designed to provide an all-in-one solution for a broad range of MCDM tasks, pymcdm stands out for its performance, modularity and user-friendly architecture. This library is built on some of the core scientific Python libraries, including NumPy for numerical operations, Matplotlib for data visualization and selected computation functions from SciPy. The

Method	Solution	CC (Min)	SS (Max)	TC (Max)	TX (Min)	TZ (Min)	MD (Max)	ML (Max)
GRA	4	1100000	5940	12	12	15	230	600
MABAC	4	1100000	5940	12	12	15	230	600
PROBID	4	1100000	5940	12	12	15	230	600
SAW	6	1500000	3465	12	6	12	260	420
TOPSIS	21	1920000	3790	10	12	20	300	1030

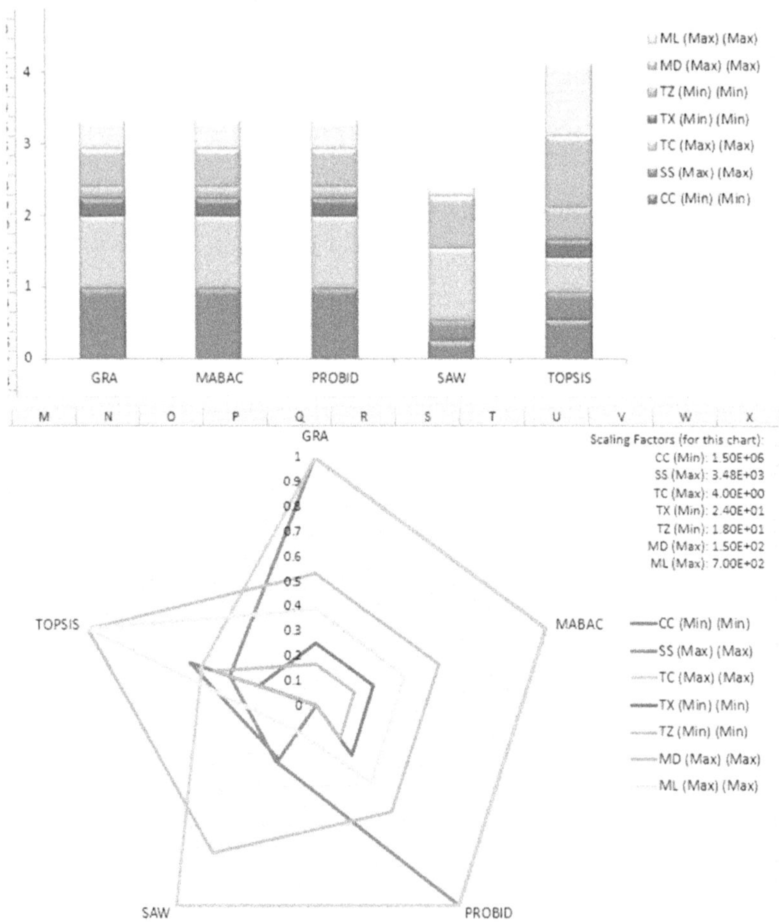

FIGURE 10.3
Selected alternatives by five different MCDM methods for an MCDM application: table, bar chart and radar diagram in the top, middle and bottom plots, respectively.

pymcdm library is structured into modular subpackages, each dedicated to core MCDM components such as normalization, weighting, ranking/MCDM, correlation analysis and visualization.

As reported by Kizielewicz et al. (2023), pymcdm Version 1.1.0 includes a total of 15 MCDM methods, namely, additive ratio assessment (ARAS), combined compromise solution (COCOSO), CODAS, characteristic objects method (COMET), COPRAS, EDAS, MABAC, multi-attribute ideal-real comparative analysis (MAIRCA), MARCOS, MOORA, operational competitiveness rating analysis (OCRA), PROMETHEE, stable preference ordering towards ideal solution (SPOTIS), TOPSIS and VIKOR. In addition, pymcdm Version 1.1.0 encompasses ten weighting methods (e.g., CRITIC, entropy, Gini and variance) and eight normalization methods (e.g., sum, vector, max–min and logarithmic). See Section 3 of Kizielewicz et al. (2023) for more details.

The pymcdm library emphasizes code readability, performance and minimal dependencies. This makes it particularly attractive to users with moderate Python experience, who require an efficient and transparent tool for both research and practical decision analysis. The pymcdm can be installed via the Python Package Index (PyPI), that is, using the "pip install pymcdm" command. In addition, the source code of pymcdm is freely available under the MIT license on GitHub (https://github.com/kotbaton/pymcdm), promoting reproducibility and community-driven development.

For readers looking to explore the pymcdm library, Kizielewicz et al. (2023) provide sample code illustrating how to use it. Alternatively, readers can use the sample code shown in Table 10.1 as a starting reference. Table 10.1 presents the code for applying the CODAS method to an example extracted from Chakraborty and Zavadskas (2014). The results have been verified to match those obtained using the CODAS implemented in EMCDM, when both tools apply the same weights generated by the EW method.

10.4 pyDecision Python Library

pyDecision is a Python-based library designed to provide a flexible, user-friendly environment for performing MCDM. Developed by Pereira et al. (2024), it is a comprehensive package that consolidates a wide array of well-known MCDM methods and weighting techniques into a single platform. The library is intended for both researchers and practitioners, who need a programmable, open-source solution for solving decision problems that involve evaluating multiple alternatives across conflicting criteria. The key objective of pyDecision is to make MCDM implementation accessible and transparent, particularly for users already familiar with Python or working

TABLE 10.1

Sample Code for Using the pymcdm Library with an Example Taken from Chakraborty and Zavadskas (2014).

```
# !pip install pymcdm # Uncomment this line for installation
import numpy as np
from pymcdm.methods import CODAS
from pymcdm.weights import entropy_weights
from pymcdm.helpers import rrankdata

# Define the alternatives-criteria matrix (ACM)
# 7 alternatives and 5 criteria
ACM = np.array([[60,   0.4,  2540,   500,  990],
                [6.35, 0.15, 1016,   3000, 1041],
                [6.8,  0.1,  1727.2, 1500, 1676],
                [10,   0.2,  1000,   2000, 965],
                [2.5,  0.1,  560,    500,  915],
                [4.5,  0.08, 1016,   350,  508],
                [3,    0.1,  1778,   1000, 920]],
                dtype='float')

# Find criteria weights using entropy weighting method
weights = entropy_weights(ACM)
# calculated weights are:
# [0.55148296, 0.13696175, 0.07917187, 0.19323723, 0.03914619]

# Define the type of each criterion,
# 1 for maximization and -1 for minimization
types = np.array ([1, -1, 1, 1, 1])

# Instantiate a CODAS object
codas = CODAS()

# Compute the performance scores and ranking of alternatives
performance_scores = codas(ACM, weights, types)
rank = rrankdata(performance_scores)

# Print and display the outcome
alternatives = [f"A{i}" for i in range(1, len(performance_scores)+1)]
print(f"{'Alternative':<11}{'Score':>8}{'Rank':>8}")
for alt, p, r in zip(alternatives, performance_scores, rank):
    print(f"{alt:<11}{p:8.2f}{int(r):8d}")

#----------output-------------
# Alternative   Score    Rank
# A1             4.94       1
# A2             0.16       2
# A3            -0.51       3
# A4            -0.61       4
# A5            -1.69       7
# A6            -1.05       5
# A7            -1.23       6
#-----------------------------
```

in data science environments. Its design prioritizes ease of use, offering clean syntax and modular functions while supporting customization and integration into larger decision support systems or optimization workflows.

As of its release in July 2025 at https://github.com/Valdecy/pyDecision, pyDecision includes over 70 MCDM methods, making it one of the most comprehensive libraries of its kind. These methods cover a wide range of methods such as CODAS, TOPSIS, VIKOR, PROMETHEE and many others. The library also supports various weighting methods, including objective techniques (e.g., CRITIC, entropy and mean) and subjective ones (e.g., AHP, BWM and rank sum), offering users flexibility in how they quantify criterion importance. In addition, pyDecision implements correlation metrics and performance indicators, allowing users to preprocess data and evaluate results from different perspectives. It also includes CR checks for AHP matrices and offers partial support for group decision-making scenarios. Visualization features, such as bar plots and radar charts, are available for displaying weights and scores of alternatives.

pyDecision is built with practicality in mind. It supports NumPy arrays and Pandas DataFrames, enabling seamless integration with data processing pipelines. It does not rely heavily on external dependencies, making it lightweight and easy to install via pip. Also, Pereira et al. (2024) provide a detailed user manual, code samples and academic references that demonstrate how the library can be used in diverse domains, ranging from supplier selection and environmental assessment to industrial project ranking. One of the notable strengths of pyDecision is its applicability in educational contexts and rapid prototyping. It allows instructors, students and analysts to model decision problems quickly and consistently, using a shared Python-based framework. The programmatic clarity of pyDecision makes it a powerful and extensible tool for decision analysis, particularly for users seeking full control over their MCDM workflows.

For readers interested in exploring the pyDecision tool, Pereira et al. (2024) have provided numerous sample codes on the library's GitHub repository (https://github.com/Valdecy/pyDecision). Both the usage examples and the full source code are freely accessible. Alternatively, readers may refer to the example presented in Table 10.2 as a starting point for using the pyDecision library. This table presents the code for applying the TOPSIS method from the pyDecision library to the example in Chakraborty and Zavadskas (2014). The results have been verified to be consistent with those produced by TOPSIS implementation in EMCDM, using the same set of ACM and weights.

If users wish to compare ranking results from multiple MCDM methods, this can be easily done in Python by importing the relevant functions and following the code example provided in Table 10.2. Table 10.3 presents an example of using the SAW and MOORA methods with the pyDecision library. Similarly, this is also applicable to the pymcdm library.

TABLE 10.2

Sample Code for Using the pyDecision Library, for an Example from Chakraborty and Zavadskas (2014).

```
# !pip install pyDecision # Uncomment this line for installation
import numpy as np
from pyDecision.algorithm import topsis_method
from pyDecision.algorithm import saw_method

# Weights
weights = [0.55148296, 0.13696175, 0.07917187, 0.19323723, 0.03914619]

# Load Criterion Type: 'max' or 'min'
types = ['max', 'min', 'max', 'max', 'max']

# ACM dataset, 7 alternatives x 5 criteria
ACM = np.array([[60,   0.4,  2540,   500,  990],
                [6.35, 0.15, 1016,   3000, 1041],
                [6.8,  0.1,  1727.2, 1500, 1676],
                [10,   0.2,  1000,   2000, 965],
                [2.5,  0.1,  560,    500,  915],
                [4.5,  0.08, 1016,   350,  508],
                [3,    0.1,  1778,   1000, 920]],
                dtype='float')

# Calculation using TOPSIS method
rank = topsis_method(ACM, criterion_type=types, weights=weights)
#--------output----------
# Alternative    Score
# A1             0.78
# A2             0.23
# A3             0.18
# A4             0.21
# A5             0.13
# A6             0.15
# A7             0.15
#-----------------------
```

10.5 RMCDA R Library

RMCDA is a comprehensive R package designed to support a diverse spectrum of MCDM methods. Recently developed by Najafi and Mirzaei (2025), it consolidates more than 50 MCDM-related methods (e.g., GRA, MABAC, TOPSIS and VIKOR). The RMCDA package can be easily installed using install.packages ("RMCDA") within the R programming environment, making it accessible to users with R programming experience. As an open-source package released under the MIT license, it offers intuitive functions and is accompanied by comprehensive documentation to support ease of use and implementation.

RMCDA is structured around modular functions that reflect the main phases of MCDM: data input, method application and result visualization.

TABLE 10.3

Sample Code of Running SAW and MOORA Methods Using the pyDecision Library.

```
from pyDecision.algorithm import saw_method
rank = saw_method(ACM, criterion_type=types, weights=weights)
#----SAW output----------
# Alternative    Score
# A1             0.713
# A2             0.381
# A3             0.362
# A4             0.329
# A5             0.204
# A6             0.244
# A7             0.278
#-----------------------

from pyDecision.algorithm import moora_method
rank = moora_method(ACM, weights, types)
#----MOORA output--------
# Alternative    Score
# A1             0.515
# A2             0.192
# A3             0.162
# A4             0.163
# A5             0.043
# A6             0.062
# A7             0.095
#-----------------------
```

Typical functions include apply.AHP(), apply.TOPSIS(), apply.PROMETHEE() and apply.SBWM(), among many others, for various methods (Najafi and Mirzaei, 2025). The package's dependencies on libraries such as dplyr, igraph, lpSolve and nloptr enable robust crafting of decision trees, flow modeling, optimization routines and interactive visualizations.

RMCDA is designed with both practitioners and analysts in mind. It supports standard data formats like CSV and DataFrames and includes user-friendly wrappers for core MCDA functionality. Moreover, RMCDA provides a web-based interface (e.g., https://najafiannice.shinyapps.io/AHP_app/), which enables nonprogrammers to upload data, choose MCDM methods and visualize results graphically, all without writing code. The package is under active development with plans to add fuzzy MCDM, advanced group decision methods and further enhancements to visualization capabilities (Najafi and Mirzaei, 2025).

For readers familiar with the R programming language, the RMCDA package is a strong option for performing MCDM. Its source code is openly available on GitHub at: https://github.com/AnniceNajafi/RMCDA. A sample code demonstrating the use of the RMCDA library is provided in Table 10.4, which demonstrates the application of the MABAC method. The results have been verified to align with those obtained from the MABAC implementation in EMCDM, using the same ACM and weights.

TABLE 10.4

Sample Code for Using the RMCDA Library, for an Example from Chakraborty and Zavadskas (2014).

```
# install.packages("RMCDA") # Uncomment this line for installation
library(RMCDA)

# ACM dataset
ACM <- matrix(
  c(
    60,   0.4,  2540,   500,  990,
    6.35, 0.15, 1016,   3000, 1041,
    6.8,  0.10, 1727.2, 1500, 1676,
    10,   0.20, 1000,   2000, 965,
    2.5,  0.10, 560,    500,  915,
    4.5,  0.08, 1016,   350,  508,
    3,    0.10, 1778,   1000, 920
   ),
  nrow = 7,
  byrow = TRUE
)
colnames(ACM) <- paste0("C", 1:5)
rownames(ACM) <- paste0("Alt", 1:7)

# Weights
w <- c(0.55148296, 0.13696175, 0.07917187, 0.19323723, 0.03914619)

# Types: 1 = maximization, -1 = minimization
types <- c(1, -1, 1, 1, 1)

# Run MABAC
pref.values <- apply.MABAC(ACM, w, types)

# Show the preference values
print(pref.values)

#--------performance scores output---------------
# Alternative     Score
# A1              0.372
# A2              0.088
# A3              0.054
# A4              0.025
# A5             -0.133
# A6             -0.111
# A7             -0.043
#-----------------------------------------------
```

10.6 SuperDecisions Desktop Software

SuperDecisions, developed by the Creative Decisions Foundation (CDF), is a popular desktop application designed to support the AHP (for both weighting and ranking) and ANP methods, both originally conceptualized by Prof. Thomas L. Saaty. As described in Chapter 8 of this book, AHP facilitates structured decision-making by decomposing complex problems into hierarchical models of goals, criteria and alternatives (Figure 8.1). ANP generalizes AHP by allowing for interdependencies and feedback among elements, providing a richer, more flexible modeling framework. SuperDecisions has been the flagship tool for applying these methodologies in real-world contexts since the late 1990s.

The genesis of SuperDecisions dates back to Saaty's pioneering work in the 1970s and 1980s, which led to the formalization of AHP (Saaty, 1990). As the theoretical foundations matured, the CDF was established to promote and support practical implementations of these methods. SuperDecisions quickly emerged as the official and most feature-rich software platform for applying AHP and ANP. Over subsequent major versions, the software has added support for network models, synthesis of priorities, sensitivity analysis and intuitive visualizations, aligning closely with evolving research and user needs.

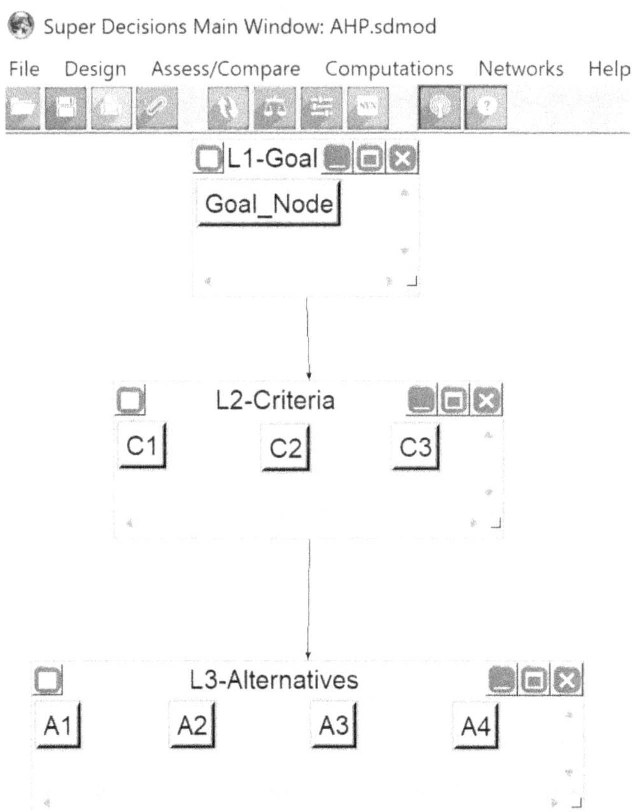

FIGURE 10.4
AHP representation in SuperDecisions software.

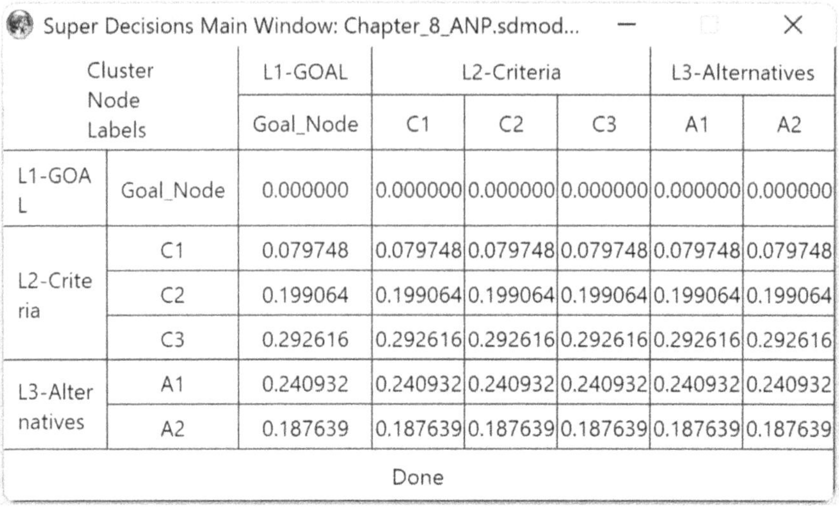

Super Decisions Main Window: Chapter_8_ANP.sdmod...

Cluster / Node / Labels		L1-GOAL	L2-Criteria			L3-Alternatives	
		Goal_Node	C1	C2	C3	A1	A2
L1-GOAL	Goal_Node	0.000000	0.000000	0.000000	0.000000	0.000000	0.000000
L2-Criteria	C1	0.079748	0.079748	0.079748	0.079748	0.079748	0.079748
	C2	0.199064	0.199064	0.199064	0.199064	0.199064	0.199064
	C3	0.292616	0.292616	0.292616	0.292616	0.292616	0.292616
L3-Alternatives	A1	0.240932	0.240932	0.240932	0.240932	0.240932	0.240932
	A2	0.187639	0.187639	0.187639	0.187639	0.187639	0.187639

Done

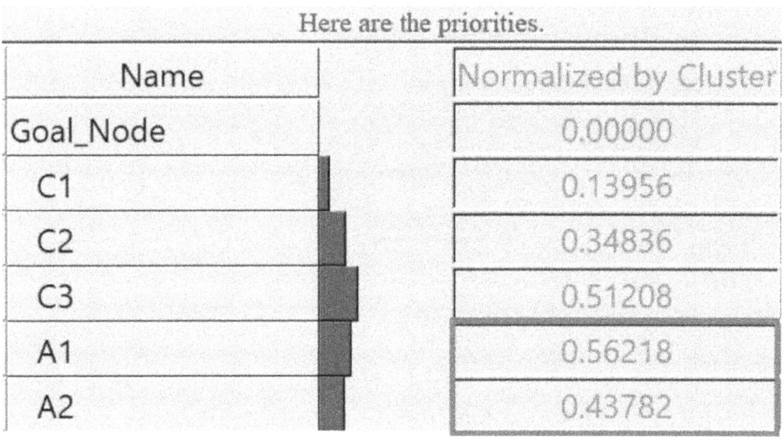

Here are the priorities.

Name	Normalized by Cluster
Goal_Node	0.00000
C1	0.13956
C2	0.34836
C3	0.51208
A1	0.56218
A2	0.43782

FIGURE 10.4 ***(Continued)***
AHP representation in SuperDecisions software.

One of the advantages of SuperDecisions is that it offers a convenient graphical user interface (GUI), such as the simplified example illustrated in Figure 10.4, which guides users seamlessly through the entire decision-making process. For a quick overview of the software, refer to the official video prepared by CDF and available at: https://youtu.be/p3nT1V0h5Po.

The use of SuperDecisions begins with model building, where users construct a hierarchical or network model by defining goals, criteria (and subcriteria, if any) and alternatives, along with the relationships between them. This intuitive modeling environment supports both the traditional AHP structure and the more complex network models required for ANP.

The software supports the entry of pairwise comparison matrices, allowing users to provide judgments between elements in a structured and systematic way. It also computes CRs to assess the logical coherence of the inputs, helping to ensure reliable results. Once these comparisons are completed, SuperDecisions proceeds to the synthesis phase, where it calculates local and global priorities. These priorities are used to rank alternatives and the software can generate sensitivity graphs to show how rankings may change in response to adjustments in weights or judgments.

For ANP models, SuperDecisions effectively manages interdependencies and feedback among elements through the construction and manipulation of a supermatrix. It performs the necessary dampening and convergence calculations to derive stable priority vectors, reflecting the influence of complex relationships within the model. The software also supports comprehensive reporting and visualization, offering interactive charts, detailed tables and export options such as spreadsheets and images for easy sharing and documentation. With its visual clarity, analytical rigor and ability to model both independent and interdependent decision structures, SuperDecisions has become a popular tool in strategic planning, policy evaluation, resource allocation and organizational decision-making contexts.

SuperDecisions is available as a desktop application and it can be downloaded from: https://www.superdecisions.com/downloads/. The official YouTube video guide for downloading the software is available at: https://youtu.be/syBobksJ6uQ. For a step-by-step tutorial on building an AHP model using SuperDecisions, refer to the official video created by the CDF at: https://youtu.be/JLs5iQrkVng.

The SuperDecisions software continues to be developed under CDF, which provides extensive user documentation, tutorials, case studies and technical support. CDF also hosts webinars and training sessions led by AHP and ANP experts, making this software both a research-grade tool and an accessible entry point for practitioners interested in using AHP and/or ANP. Recently, the CDF developed a Python library called AhpAnpLib, which can be easily installed using the pip command (i.e., pip install AhpAnpLib). The SuperDecisions website also features a dedicated section with tutorials on how to use this library at: https://www.superdecisions.com/tutorials/index.php?section=AhpAnpLib.

10.7 Other Programs/Software

In addition to the free programs, software and libraries discussed in the preceding sections, several commercial tools are also available for MCDM analysis. As these tools are not freely accessible, the authors of this book have not evaluated them directly. Readers who wish to learn more about these tools may explore them on their own. For instance, MakeItRational (https://www.softwaresuggest.com/makeitrational, accessed July 2025) is an AHP-based decision support system that enables users to define a global objective, break it down into a hierarchy of criteria and assess alternatives using either numerical values or pairwise comparisons based on Saaty's fundamental scale. The tool produces a ranked list of alternatives and supports exporting the results to Microsoft Excel.

Another example is D-Sight (https://www.d-sight.com, accessed July 2025), a commercial decision support platform that accommodates both quantitative and qualitative inputs, supports the creation of customized evaluation scales and offers interactive results based on MCDM methods such as PROMETHEE.

Additionally, the book by Papathanasiou and Ploskas (2018) provides theoretical and computational treatments of several MCDM methods along with Python code. It discusses widely used methods, such as TOPSIS, VIKOR, AHP and GP; presents extensions to these algorithms, including fuzzy theory and group decision-making; and offers step-by-step examples with Python implementations. The associated GitHub repository (https://github.com/springer-math/Multiple-Criteria-Decision-Aid) contains the source code for these implementations, allowing readers to experiment with the methods and incorporate them into their own projects. Although some of the methods are already implemented in the programs (e.g., pymcdm) described earlier in this chapter, the Python code in Papathanasiou and Ploskas (2018) provides a starting point for beginners.

10.8 Guidelines for Reliable and Robust MCDM

This section provides general guidelines for reliable and robust MCDM. It should begin by clearly defining/formulating the MCDM problem, listing all relevant alternatives and criteria and ensuring that the data behind each criterion are accurate. This first step is critical, as it requires a solid understanding of the application to include all appropriate alternatives and criteria and to ensure the criterion values are correct.

Next, explicitly tag each criterion as maximization (benefit) or minimization (cost) before any computation. If needed, transform minimization criteria into maximization form with an appropriate linear mapping (e.g., subtracting all values of a minimization criterion from a sufficiently large number). For normalization, it is generally advisable to use the original normalization equation(s) specified for each MCDM method, unless your study is to conduct a dedicated sensitivity analysis on various normalization methods (e.g., as in Baydaş et al., 2024). Furthermore, careful attention should be given to units (dimensions), outliers and missing data, as these factors can distort the validity and reliability of the decision analysis.

To enhance the robustness of results, it is recommended to try several weighting methods (e.g., based on different principles or approaches) and multiple MCDM methods (e.g., one each from the reference-, aggregation- and outranking-type methods). This multi-method strategy, known as ensemble ranking or methodology (e.g., Alves et al., 2024), helps verify whether the top-ranked alternative(s) remain consistent across different methods or not.

Additionally, sensitivity and scenario analyses are crucial for evaluating the stability of results. By systematically varying the weights, altering criteria values and/or modifying the set of alternatives (e.g., as in Nabavi et al., 2023), decision-makers can assess how such changes affect the final rankings. Visualization tools can be used to convey these dynamics effectively. Highlighting areas where the results remain invariant (or are significantly affected) provides valuable insight into the confidence and reliability of the decision recommendations (Nabavi et al., 2024).

Both ensemble methodology and sensitivity analysis are necessary, as the best weighting and ranking methods for MCDM are not yet established, which in turn is because the "correct" solution to an MCDM problem is subjective and not known. For example, Wang et al. (2022) employed five MCDM methods (GRA, MABAC, PROBID, SAW and TOPSIS) to select one Pareto-optimal solution and recommend the corresponding optimal operating condition for a supercritical water gasification process. Daftari et al. (2025) applied two weighting methods (EW and CRITIC) and five MCDM methods to the selection of the best synthesis approach of graphene oxide (GO) from among six synthesis methods. The recommended synthesis approach is based on the top ranking by the majority of these ten combinations of weighting and MCDM methods as well as the operational merits of synthesis approaches.

Finally, it is important to avoid computational errors in the application of chosen methods to the MCDM problem. To ensure error-free results, particularly for applying several weighting/ranking methods and for sensitivity analysis, select a comprehensive program/software for MCDM (e.g., one of the programs described in this chapter), familiarize with it by studying the guidelines provided and test the program on several examples (with known results, e.g., available in the literature) before using it for the (new) application.

10.9 Summary

Main points in this chapter on Programs and Software for MCDM are as follows:

1. This chapter introduced five accessible MCDM tools spanning Excel, Python, R and desktop platforms.

2. EMCDM is an Excel-based tool with built-in support for 16 MCDM and eight weighting methods via a user-friendly spreadsheet interface.
3. The pymcdm and pyDecision are Python libraries that provide extensive support for MCDM applications.
4. RMCDA is a flexible R package supporting MCDM methods through R scripts.
5. EMCDM is suitable for users without coding experience, while Python and R tools are ideal for programmatic control.
6. SuperDecisions is a stand-alone software focused on AHP and ANP, offering structured input and hierarchical modeling. It is recommended for hierarchy-based problems requiring AHP and/or ANP modeling.
7. Two commercial MCDM tools (MakeItRational and D-Sight) are also briefly mentioned in this chapter.
8. Some general guidelines (e.g., comprehensive and correct formulation of the MCDM problem, ensemble methodology, sensitivity analysis and use of a program) for reliable and robust MCDM are outlined.

References

Alves, M. A., Oliveira, B. A. S., & Guimarães, F. G. (2024). Ensemble ranking: An aggregation of multiple multicriteria methods and scenarios and its application to power generation planning. *Decision Analytics Journal*, *10*, 100435.

Baydaş, M., Kavacık, M., & Wang, Z. (2024). Interpreting the determinants of sensitivity in MCDM methods with a new perspective: An application on e-scooter selection with the probid method. *Spectrum of Engineering and Management Sciences*, *2*(1), 17–35. https://doi.org/10.31181/sems2120242b

Chakraborty, S. (2011). Applications of the MOORA method for decision making in manufacturing environment. *The International Journal of Advanced Manufacturing Technology*, *54*(9), 1155–1166.

Chakraborty, S., & Zavadskas, E. K. (2014). Applications of WASPAS method in manufacturing decision making. *Informatica*, *25*(1), 1–20.

Daftari, T., Shivanand, D., Dave, U., Nagarajan, S. L., Misra, S., Kiran, B., Rangaiah, G. P., & Chakraborty, S. (2025). Selection of optimal synthesis method of graphene oxide using multi-criteria decision-making. *Next Materials*, *8*, 100892.

Kizielewicz, B., Shekhovtsov, A., & Sałabun, W. (2023). pymcdm – the universal library for solving multi-criteria decision-making problems. *SoftwareX*, *22*, 101368.

Nabavi, S. R., Wang, Z., & Rangaiah, G. P. (2023). Sensitivity analysis of multi-criteria decision making methods for engineering applications. *Industrial & Engineering Chemistry Research*, *62*, 6707.

Nabavi, S. R., Wang, Z., & Rangaiah, G. P. (2024). Sensitivity assessment of multi-criteria decision-making methods in chemical engineering optimization applications. In F. Gómez-Castro, & V. Rico-Ramírez (Eds.), *Optimization in chemical engineering: Deterministic, meta-heuristic and data-driven techniques* (pp. 283–304). Berlin, Boston: De Gruyter. https://doi.org/10.1515/9783111383439-008

Najafi, A., & Mirzaei, S. (2025). RMCDA: The comprehensive R library for applying multi-criteria decision analysis methods. *Software Impacts, 24*, 100762.

Papathanasiou, J., & Ploskas, N. (2018). *Multiple criteria decision aid: Methods, examples and python implementation*. Springer Nature.

Pereira, V., Basilio, M. P., & Santos, C. H. T. S. H. T. (2024). Enhancing decision analysis with a large language model: Pydecision a comprehensive library of MCDA methods in python. *arXiv preprint arXiv:2404.06370*.

Saaty, T. L. (1990). How to make a decision: The analytic hierarchy process. *European Journal of Operational Research, 48*(1), 9–26.

Wang, Z., Li, J., Rangaiah, G. P., & Wu, Z. (2022). Machine learning aided multi-objective optimization and multi-criteria decision making: Framework and two applications in chemical engineering. *Computers & Chemical Engineering, 165*, 107945.

Wang, Z., Parhi, S. S., Rangaiah, G. P., & Jana, A. K. (2020). Analysis of weighting and selection methods for pareto-optimal solutions of multiobjective optimization in chemical engineering applications. *Industrial & Engineering Chemistry Research, 59*(33), 14850–14867.

Wang, Z., & Rangaiah, G. P. (2017). Application and analysis of methods for selecting an optimal solution from the Pareto-optimal front obtained by multiobjective optimization. *Industrial & Engineering Chemistry Research, 56*(2), 560–574.

Wang, Z., Rangaiah, G. P., & Wang, X. (2021). Preference ranking on the basis of ideal-average distance method for multi-criteria decision-making. *Industrial & Engineering Chemistry Research, 60*(30), 11216–11230.

Exercises

E10.1 You are given an ACM consisting of three alternatives (A1, A2 and A3) and four criteria (C1, C2, C3 and C4), with the following values:

Alternatives	C1 (Max)	C2 (Max)	C3 (Max)	C4 (Min)
A1	4.21	185	0.071	10.8
A2	5.98	820	0.056	13.9
A3	7.77	391	0.031	17.2

Criteria C1, C2 and C3 are benefit criteria to be maximized, whereas C4 is a cost criterion to be minimized. Select one of the weighting methods available in the EMCDM program and apply it to run all the implemented MCDM methods. Identify the top-ranked alternative from each method and compare the results. Then, change the weighting method to another and execute all the MCDM methods in EMCDM. Analyze the ranking results for the effect of the weighting method and the MCDM method.

E10.2 You are given an ACM consisting of six alternatives (A1, A2, A3, A4, A5 and A6) and three criteria (C1, C2 and C3), with the following values:

Alternatives	C1 (Max)	C2 (Max)	C3 (Min)
A1	365.9	0.869	98.6
A2	915.8	0.582	84.3
A3	575.7	0.193	37.4
A4	351.2	0.815	75.5
A5	614.3	0.104	70.0
A6	698.5	0.265	77.2

Criteria C1 and C2 are benefit criteria to be maximized, whereas C3 is a cost criterion to be minimized. Using the pymcdm library, first apply the entropy method to determine the criteria weights. Then, select five MCDM methods available in the library and write Python code to execute them. Determine the top-ranked alternative for each method and compare the results.

E10.3 You are given an ACM consisting of four alternatives (A1, A2, A3 and A4) and five criteria (C1, C2, C3, C4 and C5), with the following values:

Alternatives	C1 (Max)	C2 (Max)	C3 (Min)	C4 (Max)	C5 (Min)
A1	64.6	0.020	9.73	30.28	0.088
A2	81.0	0.031	2.38	13.90	0.060
A3	77.3	0.125	10.13	23.94	0.035
A4	61.2	0.255	2.68	40.77	0.013

Criteria C1, C2 and C4 are benefit criteria to be maximized, whereas C3 and C5 are cost criteria to be minimized. Using the pyDecision library, first apply the CRITIC method to determine the criteria weights. Then, select five MCDM methods available in the library and write Python code to execute them. Determine the top-ranked alternative for each method and compare the results.

E10.4 You are given an ACM consisting of five alternatives (A1, A2, A3, A4 and A5) and four criteria (C1, C2, C3 and C4), with the following values:

Alternatives	C1 (Max)	C2 (Max)	C3 (Min)	C4 (Min)
A1	4521.6	3.954	0.025	0.007
A2	1819.5	4.141	0.128	0.121
A3	753.7	8.964	0.059	0.063
A4	5044.7	7.328	0.003	0.072
A5	1507.2	2.130	0.025	0.125

Criteria C1 and C2 are benefit criteria to be maximized, whereas C3 and C4 are cost criteria to be minimized. Using the RMCDA library, first apply the entropy method to determine the criteria weights. Then, select five MCDM methods available in the library and write Python code to execute them. Determine the top-ranked alternative for each method and compare the results.

E10.5 Use the AHP method in the SuperDecisions software to determine the subjective weights of the four criteria from Exercise E10.4 and then continue using AHP and ANP to rank the five alternatives. You may provide all your own pairwise comparison judgments as part of the assessment.

E10.6 Which of the programs/software covered in this chapter will you choose for your application, considering its nature and requirements? Give reason(s) for your choice. In case none of them is suitable, state the reason(s).

Appendix: Alternatives–Criteria Matrix (ACM) for Selected Applications

Overview

In multi-criteria decision-making (MCDM), alternatives-criteria matrix (ACM) data are central and essential. Further, its development (i.e., finding all relevant alternatives, criteria and their values) through surveys of experts, markets or literature, design and analysis of systems, or solution of MOO problems is equally important and often requires substantial effort.

The purpose of this appendix is to provide many sets of ACM data for carefully chosen diverse applications from the open literature. Readers can use any of them to apply the methods and programs described in Chapters 3–10 as well as to help them in finding relevant and right ACM data for their own application(s). This appendix contains 26 applications and they are grouped into scientific (three applications), engineering (nine applications), management (eight applications) and finance (six applications) categories. This grouping is purely for convenience and it is possible that some of these applications can be shifted from one category to another.

In the following, brief description, criteria and reference for each application are given, whereas ACM data are provided in a Microsoft Excel file (accessible from the shared Google Drive at https://drive.google.com/drive/folders/18ynIwnKaCRsptwBCrMIFVOlV9Mu_A7qM) so that readers can readily use the ACM data with any program of their choice and for conciseness in the book. ACM data for applications are presented in the Excel file in an easy-to-use format.

ACM Data for Scientific Applications

Biomass Selection: Biomass is a renewable resource for producing fuels, chemicals and other products. The composition of available biomasses such as rice and wheat straw, sugarcane bagasse and palm shells varies. Madhu et al. (2020) applied MCDM to select one of the seven biomasses (alternatives) considering seven criteria related to composition. The following criteria are weight percentages: C1 (Min) – lignin content, C2 (Max) – cellulose content, C3 (Max) – hemicellulose content, C4 (Max) – volatile matter, C5 (Max) – fixed carbon, C6 (Min) – moisture content and C7 (Min) – ash content.

Reinforced Composite Design: Singh (2021) studied MCDM application to the optimal design of reinforced composite for automotive brakes; he measured values of seven criteria related to tribological properties for each composite. The ACM data have 12 alternatives (i.e., composites of different percentages of materials). The seven criteria are: C1 (Max) – performance coefficient of friction, C2 (Min) – friction fluctuations, C3 (Max) – friction stability, C4 (Min) – friction variability, C5 (Min) – friction fade, C6 (Max) – friction recovery and C7 (Min) – wear.

Graphene Oxide (GO) Synthesis Method Selection: In the literature, there are many synthesis methods for GO synthesis, with each having its own advantages and disadvantages. Daftari et al. (2025) applied MCDM to the selection of the best method among the six synthesis methods (alternatives). The six criteria employed are: C1 (Min) – toxicity, C2 (Min) – maximum temperature in the synthesis process, C3 (Max) – minimum temperature in the synthesis process, C4 (Max) – ratio of mass of GO to carbon source (yield), C5 (Max) – ratio of mass of GO produced to acid used and C6 (Max) – ratio of mass of GO produced to oxidizer used.

ACM Data for Engineering Applications

Forklift Selection: Some details of this application are given in Table 1.1 and associated text in Chapter 1. ACM data provided in the MS Excel file are taken from Table 7 (Pamučar and Ćirović, 2015), who also describe the development of these data. This MCDM application involves ten criteria and seven alternatives (i.e., forklift models available in the market). The criteria are: C1 (Min) – purchase price, C2 (Max) – manufacturer's warranty period (years), C3 (Max) – service network, C4 (Max) – availability of spare parts, C5 (Min) – average cost of maintenance, C6 (Min) – fuel consumption, C7 (Max) – maximum bearing capacity, C8 (Max) – maximum lifting capacity, C9 (Max) – movement speed of the forklift with/without load and C10 (Max) – speed of lifting/lowering loads.

Computerized Numerical Control (CNC) Machine Selection: Chakraborty (2011) evaluated an MCDM method for this and many other applications from the literature. He reported ACM data for this application as machine tool selection in Table 5, which is based on Sun (2002). It consists of 21 alternatives (i.e., different models of CNC machines). The seven criteria are: C1 (Min) – capital cost, C2 (Max) – spindle speed range, C3 (Max) – tool capacity, C4 (Min) – rapid traverse rate along the x-axis, C5 (Min) – rapid traverse rate along the z-axis, C6 (Max) – maximum machining diameter and C7 (Max) – maximum machining length.

Industrial Robot Selection: Bhangale et al. (2004) presented an example of selecting a robot for some pick-n-place operation from seven candidate robots (alternatives). This robot selection involves five criteria, which are: C1 (Max) – load capacity (kg), C2 (Min) – repeatability (mm), C3 (Max) – maximum tip speed (mm/s), C4 (Max) – memory capacity (points or steps) and C5 (Max) – manipulator reach (mm).

Volatile Organic Components Recovery Process Design: Some details of this application are given in Table 1.1 and associated text in Chapter 1. The ACM data provided in the MS Excel file were obtained from the multi-objective optimization (MOO) of the design of a process for recovering volatile organic components by Lee and Rangaiah (2009); it is for Case F in Table 2 in this article. This MCDM application, used as VOC5 in Nabavi et al. (2023), has five criteria and 74 alternatives or Pareto-optimal solutions (having positive values for all five criteria). The criteria are: C1 (Max) – net present value, C2 (Min) – human toxicity potential, C3 (Min) – ecotoxicity potential, C4 (Min) – atmospheric potential and C5 (Min) – photochemical oxidation potential.

External Wall Finishing in Cast-in-Place Buildings: ACM data for this application are taken from Zavadskas and Turskis (2008). It is for five technological alternatives for a 13-story block of flats, which were evaluated for four criteria by 28 experts. The criteria are: C1 (Min) – cost in thousands, C2 (Min) – labor input in man-days, C3 (Max) – durability and C4 (Max) – aesthetics.

Landing Site for Drones: This application from Vafaei et al. (2018) is on the selection of a site from 16 alternative sites for the landing of drones, avoiding hazards. The three criteria are: C1 (Max) – illumination; C2 (Max) – reachability and C3 (Min) – land texture.

Passive Design Strategy for High-Performance Buildings: In this application, Rachman et al. (2024) analyzed passive design strategies for high-performance buildings through MOO for four criteria and report ACM data from the obtained Pareto-optimal solutions. There are 11 alternatives. The criteria are: C1 (Min) – energy use intensity, C2 (Max) – percentage of comfort hours, C3 (Max) – spatial daylight autonomy and C4 (Min) – annual sunlight exposure.

Renewable Energy Sources in Taiwan: This application on the selection of a renewable energy source from six alternatives available in Taiwan, was analyzed by Lee and Chang (2018). It considers ten criteria, which are: C1 (Min) – investment cost for the power plant, C2 (Min) – operation and maintenance cost, C3 (Min) – electricity cost, C4 (Max) – efficiency (ratio of output energy to input energy, expressed as a percent), C5 (Max) – capacity factor (ratio of energy produced to installed capacity, expressed as a percent), C6 (Max) – technical maturity (i.e., reliability of the technology), C7 (Min) – greenhouse gas emissions, C8 (Min) – land required for energy production, C9 (Max) – number of jobs created and C10 (Max) – social acceptance of renewable energy source.

Renewable Energy Sources in India: Husain et al. (2024) investigated the selection of the best renewable energy source among four alternatives (namely, solar, wind, hydro and biomass) in the context of India. Like Lee and Chang (2018), they too considered ten criteria, which are: C1 (Min) – capital cost ($/kW), C2 (Min) – operation and maintenance cost ($/kW.y), C3 (Min) – levelized cost of electricity ($/kWh), C4 (Max) – efficiency (%), C5 (Max) – capacity factor (%), C6 (Min) – greenhouse gas emission (gCO2/kWh), C7 (Min) – land requirement (m2/kW), C8 (Max) – jobs creation (job-years/GWh), C9 (Max) – technical maturity and C10 (Max) – social acceptance. Both C9 and C10 are on a scale of 1–5.

ACM Data for Management Applications

Location for a Logistical Center: This application is for selecting the best alternative from eight possible locations for a logistical center, considering 11 criteria. ACM data are taken from Table 3 in Mukhametzyanov and Pamučar (2018). The criteria are: C1 (Max) – connectivity to multimodal transport, C2 (Max) – infrastructure development evaluation, C3 (Min) – environmental impact, C4 (Max) – conformity with spatial plans and strategy of economic development, C5 (Max) – gravitating intermodal transport unit, C6 (Max) – reload capacity of the center, C7 (Max) – available area for future development and capacity expansion of the center, C8 (Min) – user's distance from the center; C9 (Max) – traffic safety, C10 (Max) – length of railroad reload front and C11 (Max) – quality of traffic approaches for interpellant means of transportation.

Agricultural Distribution Center: Puska et al. (2024) reported ACM data for selecting a location for an agricultural distribution center from among ten locations, considering ten criteria. The criteria are: C1 (Max) – possibility of multimodal transport, C2 (Max) – size and expandability, C3 (Min) – distance to customers and suppliers, C4 (Min) – land costs, C5 (Min) – logistics costs, C6 (Min) – distance to airport, C7 (Min) – distance to highway, C8 (Min) – distance to railway, C9 (Max) – developed infrastructure and C10 (Min) – environmental impact.

Fruit Supply Chain: Cheraghalipour et al. (2018) investigated the MOO of a closed-loop fruit supply chain for two objectives. Based on the Pareto-optimal solutions reported in this work, Wang et al. (2023) used ACM data that have 16 alternatives and two criteria: C1 (Min) – total cost and C2 (Max) – satisfaction of the demand (termed as responsiveness).

Highway Construction Scheduling: Zhao et al. (2025) investigated MOO of low-carbon highway construction scheduling for three criteria (objectives) and reported 30 Pareto-optimal solutions (alternatives), which constitute ACM data for MCDM. The criteria are: C1 (Min) – total duration (days), C2 (Min) – total cost and C3 (min) – carbon emissions (kg CO_2).

Energy Management System for a Desalination Plant: In this application, Rezk et al. (2021) employed MCDM for choosing an energy management system for a desalination plant based on reverse osmosis (RO). They considered four energy management strategies for supplying energy from a photovoltaic–diesel–battery system to three sizes of the RO plant. In all, there are 12 alternatives and their performance is evaluated by eight criteria related to energy consumption and environmental effects. The criteria are: C1 (Min) – operating cost ($/year), C2 (Max) – renewable fraction, C3 (Min) – initial cost ($), C4 (Min) – cost of energy ($/kWh), C5 (Min) – excess energy (kWh), C6 (Min) – unmet load (kWh), C7 (Min) – breakeven grid extension distance (km) and C8 (Min) – carbon dioxide emissions (kg/year).

Selection of Airport Location in Poland: The ACM data for this application, reported by Palczewski and Salabun (2019), consist of seven criteria and ten alternatives for airport location. The criteria are: C1 (Max) – population within 100 km range, C2 (Min) – distance from the biggest city in the region, C3 (Max) – available infrastructure (e.g., roads and electric power), C4 (Max) – distance from nature reserves and protection zones (to minimize harmful effects on the surrounding environment), C5 (Min) – geological structure of the location's terrain and its surroundings, C6 (Min) – time to reach the airport from the city center by public transport and C7 (Min) – investment for building the airport.

Industrial Economy of Provinces/Regions: In this application studied by Wang and Luo (2010), the goal is to assess the economic benefits of industries in different provinces (regions) in China. The ACM data have five criteria and 16 alternatives (provinces). The criteria are: C1 (Max) – overall labor productivity of industries, C2 (Max) – ratio of profits and total taxes to the average value of total assets, C3 (Max) – gross profits from sales revenue, C4 (Min) – ratio of current fund employed to industrial output value and C5 (Max) – ratio of total pretax profits to the gross output value.

Ranking Educational Institutions: Das et al. (2012) studied the performance evaluation of seven Indian Institute of Technology (IITs) in terms of six criteria, which are: C1 (Max) – faculty strength, C2

(Max) – student intake, C3 (Max) – number of Ph.D.s awarded, C4 (Max) – number of patents applied for, C5 (Max) – campus area and C6 (Min) – tuition fee per semester.

ACM Data for Finance Applications

Listed Firms in Borsa İstanbul (BIST): Baydaş and Pamučar (2022) presented ACM data for evaluating and ranking the financial performance of 24 publicly traded companies listed on the BIST-30 index, based on six financial criteria. The criteria are: C1 (Max) – Altman Z-score, indicating bankruptcy risk and financial health; C2 (Max) – return on equity (ROE), measuring profitability relative to shareholder equity; C3 (Max) – return on assets (ROA), assessing efficiency in asset utilization; C4 (Max) – market value added margin (MVA Margin), reflecting value creation per unit of sales; C5 (Max) – MVA Spread, capturing value generation relative to invested capital; and C6 (Max) – market-to-book value ratio (MV/BV), representing investor valuation of the company's equity.

European Financial Institutions: Ragazou et al. (2025) showcased ACM data for ranking and assessing eco-efficient European financial institutions from among 52 alternatives, considering ten criteria. The criteria are: C1 (Max) – emissions score, C2 (Max) – policy energy efficiency score, C3 (Max) – targets water efficiency score, C4 (Max) – policy business ethics score, C5 (Max) – policy fair competition score, C6 (Max) – corporate social responsibility (CSR) reporting score, C7 (Max) – human rights score, C8 (Max) – board gender diversity (percent) score, C9 (Max) – customer satisfaction score and C10 (Max) – policy labor score.

Development and Investment Banks: ACM data for assessing and ranking the financial performance of 11 development and investment banks were reported in Erdoğan (2023). The analysis considered 11 criteria: C1 (Max) – total assets, C2 (Max) – shareholders' equity-to-total assets ratio, C3 (Min) – nonperforming loans-to-total loans ratio, C4 (Min) – fixed assets-to-total assets ratio, C5 (Max) – liquid assets-to-total assets ratio, C6 (Max) – liquid assets-to-short-term liabilities ratio, C7 (Max) – average ROA, C8 (Max) – average return on shareholders' equity, C9 (Min) – other operating expenses-to-operating gross profit ratio, C10 (Max) – operating gross profit-to-total assets ratio and C11 (Max) – net operating profit (loss)-to-total assets ratio.

Financial Performance Evaluation: Yang (2022) utilized ACM data for ranking and evaluating the financial performance of 20 publicly listed manufacturing firms based in the United States, using data from their 2021 financial statements. The analysis considered 12 financial criteria: C1 (Max) – current ratio, C2 (Max) – acid test ratio, C3 (Max) – total asset turnover ratio, C4 (Max) – inventory turnover ratio, C5 (Max) – equity ratio, C6 (Min) – debt equity ratio, C7 (Min) – debt-to-total capital ratio, C8 (Max) – net profit margin, C9 (Max) – return on net capital, C10 (Max) – operating cash flow ratio, C11 (Min) – fixed asset-to-net worth ratio and C12 (Max) – diluted earnings per share (EPS).

Financial and Diversity Performance: Bouslah et al. (2023) presented ACM data for ranking and evaluating the financial and diversity performance of 20 publicly listed firms. The analysis considered eight criteria: C1 (Max) – ROA, C2 (Max) – Tobin's Q, C3 (Max) – board cultural diversity, C4 (Max) – percentage of women employees, C5 (Max) – percentage of new women employees, C6 (Max) – percentage of women managers, C7 (Max) – percentage of females on the board and C8 (Max) – percentage of female executive members.

Portfolio Performance Models: Sikalo et al. (2023) reported ACM data for ranking and evaluating the performance of six portfolio optimization models using weekly trading data from 57 stocks listed on the STOXX Europe 600 Index over the period 2000–2020. The analysis considered 11 criteria: C1 (Min) – SD, C2 (Min) – mean absolute deviation (MAD), C3 (Min) – conditional value at risk (CVaR), C4 (Max) – maximum loss (ML), C5 (Max) – average weekly return (AR), C6 (Max) – cumulative return (CR), C7 (Max) – Sharpe ratio (SR), C8 (Max) – Calmar ratio (CMR), C9 (Max) – Omega ratio (OR), C10 (Max) – CVaR ratio (CVR) and C11 (Max) – predictability (PRED). These criteria capture four dimensions of portfolio performance: risk (SD, MAD, CVaR and ML), return (AR and CR), stability/return-to-risk ratios (SR, CMR, OR and CVR) and predictability (PRED).

References

Baydaş, M., & Pamučar, D. (2022). Determining objective characteristics of MCDM methods under uncertainty: An exploration study with financial data. *Mathematics, 10*(7), 1115.

Bhangale, P., Agrawal, V. P., & Saha, S. K. (2004). Attribute based specification, comparison and selection of a robot. *Mechanism and Machine Theory, 39*, 1345–1366.

Bouslah, K., Liern, V., Ouenniche, J., & Pérez-Gladish, B. (2023). Ranking firms based on their financial and diversity performance using multiple-stage unweighted TOPSIS. *International Transactions in Operational Research, 30*(5), 2485–2505.

Chakraborty, S. (2011). Applications of the MOORA method for decision making in manufacturing environment. *International Journal of Advanced Manufacturing Technology, 54*, 1155–1166.

Cheraghalipour, A., Paydar, M. M., & Hajiaghaei-Keshteli, M. (2018). A bi-objective optimization for citrus closed-loop supply chain using Pareto-based algorithms. *Applied Soft Computing, 69*, 33–59.

Daftari, T., Shivanand, D., Dave, U., Nagarajan, S. L., Misra, S., Kiran, B., Rangaiah, G. P., & Chakraborty, S. (2025). Selection of optimal synthesis method of graphene oxide using multi-criteria decision-making. *Next Materials, 8*, 100892.

Das, M. C., Sarkar, B., & Ray, S. (2012). A framework to measure relative performance of Indian technical institutions using integrated fuzzy AHP and COPRAS methodology. *Socio-Economic Planning Sciences, 46*, 230–241.

Erdoğan, B. (2023). Financial performance analysis of development and investment banks: TOPSIS method. *Dicle Üniversitesi Sosyal Bilimler Enstitüsü Dergisi*, (34), 1–15.

Husain, A. M., Hasan, M. M., Khan, Z. A., & Asjad, M. (2024). A robust decision-making approach for the selection of an optimal renewable energy source in India. *Energy Conversion and Management, 301*, 117989.

Lee, H. C., & Chang, C. T. (2018). Comparative analysis of MCDM methods for ranking renewable energy sources in Taiwan. *Renewable and Sustainable Energy Reviews, 92*, 883–896.

Lee, S. Q. E., & Rangaiah, G. P. (2009). Optimization of recovery processes for multiple economic and environmental objectives. *Industrial and Engineering Chemistry Research, 48*, 7662–7681.

Madhu, P., Dhanalakshmi, C. S., & Mathew, M. (2020). Multi-criteria decision-making in the selection of a suitable biomass material for maximum bio-oil yield during pyrolysis. *Fuel, 277*, 118109.

Mukhametzyanov, I., & Pamučar, D. (2018). A sensitivity analysis in MCDM problems: A statistical approach. *Decision Making: Applications in Management and Engineering, 1*(2), 51–80.

Nabavi, S. R., Wang, Z., & Rangaiah, G. P. (2023). *Sensitivity Analysis of Multi-Criteria Decision Making Methods for Engineering Applications, Industrial & Engineering Chemistry Research, 62*, 6707.

Palczewski, K., & Salabun, W. (2019). Influence of various normalization methods in PROMETHEE II: An empirical study on the selection of the airport location. *Procedia Computer Science, 159*, 2051–2060.

Pamučar, D., & Ćirović, G. (2015). The selection of transport and handling resources in logistics centers using Multi-Attributive Border Approximation Area Comparison (MABAC). *Expert Systems with Applications, 42*, 3016.

Puska, A., Stilic, A., Pamučar, D., Bozanic, D., & Nedeljkovic, M. (2024). Introducing a novel multi-criteria Ranking of Alternatives with Weights of Criterion (RAWEC) model. *MethodsX, 12*, 102628.

Rachman, A. P., Ichwania, C., Mangkuto, R. A., Pradipta, J., Koerniawan, M. D., & Sarwono, J. (2024). Comparison of multi-criteria decision making methods for selection of optimum passive design strategy. *Energy and Buildings, 314*, 114285.

Ragazou, K., Lemonakis, C., Passas, I., Zopounidis, C., & Garefalakis, A. (2025). ESG-driven ecopreneur selection in European financial institutions: Entropy and TOPSIS analysis. *Management Decision, 63*(4), 1316–1345.

Rezk, H., Alamri, B., Aly, M., Fathy, A., Olabi, A. G., Abdelkareem, M. A., & Ziedan, H. A. (2021). Multicriteria decision-making to determine the optimal energy management strategy of hybrid PV-diesel battery-based desalination system. *Sustainability, 13*, 4202.

Sikalo, M., Arnaut-Berilo, A., & Delalic, A. (2023). A combined AHP-PROMETHEE approach for portfolio performance comparison. *International Journal of Financial Studies, 11*(1), 46.

Singh, T. (2021). Optimum design based on fabricated natural fiber reinforced automotive brake friction composites using hybrid CRITIC-MEW approach. *Journal of Materials Research and Technology*, *14*, 81–92.

Sun, S. (2002). Assessing computer numerical control machines using data envelopment analysis. *International Journal of Production Research*, *40*(9), 2011–2039.

Vafaei, N., Ribeiro, R. A., & Camarinha-Matos, L. M. (2018). Data normalisation techniques in decision making: Case study with TOPSIS method. *International Journal of Information and Decision Sciences*, *10*(1), 19–38.

Wang, Y. M., & Luo, Y. (2010). Integration of correlations with standard deviations for determining attribute weights in multiple attribute decision making. *Mathematical and Computer Modeling*, *51*, 1–12.

Wang, Z., Nabavi, S. R., & Rangaiah, G. P. (2023). Selected multi-criteria decision-making methods and their applications to product and system design. In *Optimization methods for product and system design* (pp. 107–138). Singapore: Springer Nature Singapore.

Yang, K. (2022). Application of TOPSIS technique for financial performance evaluation of manufacturing firms in US market. *Academic Journal of Business and Management*, *4*(16), 116–122.

Zavadskas, E. K., & Turskis, Z. (2008). A new logarithmic normalization method in games theory. *Informatica*, *19*(2), 203–314.

Zhao, N., Zhang, J., Ding, C., & Zhang, J. (2025). Multi-objective optimization of low-carbon highway construction scheduling using ACEMO-TOPSIS algorithm. *Journal of Cleaner Production*, *522*, 146371.

Index

D

E

F

G

H

I

L

M

N

T

U

V

W